Variational Methods for Free Surface Interfaces

Variational Methods for Free Surface Interfaces

Proceedings of a Conference Held at Vallombrosa Center,
Menlo Park, California, September 7–12, 1985

Edited by
Paul Concus and Robert Finn

Organizing Committee
R. Brown, Massachusetts Institute of Technology
P. Concus, University of California, Berkeley
R. Finn (Chairman), Stanford University
S. Hildebrandt, University of Bonn
M. Miranda, University of Trento

With 44 Figures

Springer-Verlag
New York Berlin Heidelberg
London Paris Tokyo

Paul Concus
Lawrence Berkeley Laboratory
and Department of Mathematics
University of California
Berkeley, California 94720
U.S.A.

Robert Finn
Department of Mathematics
Stanford University
Stanford, California 94305
U.S.A.

AMS Classification: 49F10, 35-06, 53A10

Library of Congress Cataloging-in-Publication Data
Variational methods for free surface interfaces.
  Bibliography: p.
  1. Surfaces (Technology)—Congresses.   2. Surfaces—
Congresses.   3. Surface chemistry—Congresses.
I. Concus, Paul.   II. Finn, Robert.
TA407.V27   1986   620.1'129   86-27899

Typeset by Asco Trade Typesetting Ltd., Hong Kong.
Printed and bound by R.R. Donnelley and Sons, Harrisonburg, Virginia.
Printed in the United States of America.

9 8 7 6 5 4 3 2 1

ISBN 0-387-96396-0 Springer-Verlag New York Berlin Heidelberg
ISBN 3-540-96396-0 Springer-Verlag Berlin Heidelberg New York

Preface

Vallombrosa Center was host during the week September 7–12, 1985 to about 40 mathematicians, physical scientists, and engineers, who share a common interest in free surface phenomena. This volume includes a selection of contributions by participants and also a few papers by interested scientists who were unable to attend in person. Although a proceedings volume cannot recapture entirely the stimulus of personal interaction that ultimately is the best justification for such a gathering, we do offer what we hope is a representative sampling of the contributions, indicating something of the varied and interrelated ways with which these classical but largely unsettled questions are currently being attacked.

For the participants, and also for other specialists, the 23 papers that follow should help to establish and to maintain the new ideas and insights that were presented, as active working tools. Much of the material will certainly be of interest also for a broader audience, as it impinges and overlaps with varying directions of scientific development.

On behalf of the organizing committee, we thank the speakers for excellent, well-prepared lectures. Additionally, the many lively informal discussions did much to contribute to the success of the conference.

The participants benefited greatly from the warm and pleasant ambience provided by the Vallombrosa Center and its friendly and helpful staff, to whom we wish to offer our special thanks. The conference was made possible in part by support from the Air Force Office of Scientific Research, the Department of Energy, the National Science Foundation, and the Office of Naval Research. The National Science Foundation served as coordinating agency.

Paul Concus
Robert Finn

Contents

Preface v

List of Contributors ix

Optimal Crystal Shapes
JEAN E. TAYLOR and F.J. ALMGREN, JR. 1

Immersed Tori of Constant Mean Curvature in R^3
HENRY C. WENTE 13

The Construction of Families of Embedded Minimal Surfaces
DAVID A. HOFFMAN 27

Boundary Behavior of Nonparametric Minimal Surfaces—Some Theorems
and Conjectures
KIRK E. LANCASTER 37

On Two Isoperimetric Problems with Free Boundary Conditions
S. HILDEBRANDT 43

Free Boundary Problems for Surfaces of Constant Mean Curvature
MICHAEL STRUWE 53

On the Existence of Embedded Minimal Surfaces of Higher Genus with Free
Boundaries in Riemannian Manifolds
JÜRGEN JOST 65

Free Boundaries in Geometric Measure Theory and Applications
MICHAEL GRÜTER 77

A Mathematical Description of Equilibrium Surfaces
MARIO MIRANDA.. 85

Interfaces of Prescribed Mean Curvature
I. TAMANINI.. 91

On the Uniqueness of Capillary Surfaces
LUEN-FAI TAM 99

The Behavior of a Capillary Surface for Small Bond Number
DAVID SIEGEL... 109

Convexity Properties of Solutions to Elliptic P.D.E.'s
NICHOLAS J. KOREVAAR 115

Boundary Behavior of Capillary Surfaces via the Maximum Principle
GARY M. LIEBERMAN... 123

Convex Functions Methods in the Dirichlet Problem for Euler–Lagrange
Equations
ILYA J. BAKELMAN... 127

Stability of a Drop Trapped Between Two Parallel Planes: Preliminary Report
THOMAS I. VOGEL 139

The Limit of Stability of Axisymmetric Rotating Drops
FREDERIC BRULOIS 145

Numerical Methods for Propagating Fronts
JAMES A. SETHIAN 155

A Dynamic Free Surface Deformation Driven by Anisotropic Interfacial
Forces
DANIEL ZINEMANAS and AVINOAM NIR 165

Stationary Flows in Viscous Fluid Bodies
JOSEF BEMELMANS 173

Large Time Behavior for the Solution of the Non-Steady Dam Problem
DIETMAR KRÖNER... 179

New Results Concerning the Singular Solutions of the Capillarity Equation
MARIE-FRANÇOISE BIDAUT-VERON 191

Continuous and Discontinuous Disappearance of Capillary Surfaces
PAUL CONCUS and ROBERT FINN... 197

List of Contributors

F.J. ALMGREN, JR., Mathematics Department, Princeton University, Princeton, New Jersey 08903, U.S.A.

ILYA J. BAKELMAN, Department of Mathematics, Texas A&M University, College Station, Texas 77843, U.S.A.

JOSEF BEMELMANS, Fachbereich Mathematik, Universität des Saarlandes, 6600 Saarbrücken, Federal Republic of Germany

MARIE-FRANÇOISE BIDAUT-VERON, Department of Mathematics, University of Tours, 37200 Tours, France

FREDERIC BRULOIS, California State University, Dominguez Hills, Carson, California 90747, U.S.A.

PAUL CONCUS, Lawrence Berkeley Laboratory and Department of Mathematics, University of California, Berkeley, California 94720, U.S.A.

ROBERT FINN, Department of Mathematics, Stanford University, Stanford, California 94305, U.S.A.

MICHAEL GRÜTER, University of Bonn, Mathematics Institute, 5300 Bonn, Federal Republic of Germany

S. HILDEBRANDT, University of Bonn, Mathematics Institute, 5300 Bonn, Federal Republic of Germany

DAVID A. HOFFMAN, Department of Mathematics, University of Massachusetts, Amherst, Massachusetts 01003, U.S.A.

JÜRGEN JOST, University of Bochum, Mathematics Institute, 4630 Bochum-Querenburg, Federal Republic of Germany

NICHOLAS J. KOREVAAR, Department of Mathematics, University of Kentucky, Lexington, Kentucky 40506, U.S.A.

DIETMAR KRÖNER, University of Bonn, Mathematics Institute, 5300 Bonn, Federal Republic of Germany

KIRK E. LANCASTER, Department of Mathematics and Statistics, Wichita State University, Wichita, Kansas 67208, U.S.A.

GARY M. LIEBERMAN, Department of Mathematics, Iowa State University, Ames, Iowa 50011, U.S.A.

MARIO MIRANDA, Institute of Mathematics, University of Trento, 38100 Trento, Italy

AVINOAM NIR, Department of Chemical Engineering, Technion, Haifa 32000, Israel

JAMES A. SETHIAN, Lawrence Berkeley Laboratory and Department of Mathematics, University of California, Berkeley, California 94720, U.S.A.

DAVID SIEGEL, Department of Applied Mathematics, University of Waterloo, Waterloo, Ontario N2L 3G1 Canada

MICHAEL STRUWE, ETH-Zentrum, Mathematics Institute, CH-8092 Zürich, Switzerland

LUEN-FAI TAM, Department of Mathematics, University of Illinois, Chicago, Illinois 60680, U.S.A.

I. TAMANINI, University of Trento, Department of Mathematics, 38050 Trento, Italy

JEAN E. TAYLOR, Mathematics Department, Rutgers University, New Brunswick, New Jersey 08903, U.S.A.

THOMAS I. VOGEL, Department of Mathematics, Texas A&M University, College Station, Texas 77843, U.S.A.

HENRY C. WENTE, Department of Mathematics, University of Toledo, Toledo, Ohio 43606, U.S.A.

DANIEL ZINEMANAS, Department of Chemical Engineering, Technion, Haifa 32000, Israel

Optimal Crystal Shapes

Jean E. Taylor and F.J. Almgren, Jr.

1. Introduction

Associated with any Borel function Φ defined on the unit sphere S^n in R^{n+1} with values in $R \cup \{\infty\}$ (and, say, bounded from below) and any n-dimensional oriented rectifiable surface S in R^{n+1} is the integral

$$\Phi(S) = \int_{x \in S} \Phi(v_S(x)) \, dH^n x;$$

here $v_S(\cdot)$ denotes the unit normal vectorfield orienting S, and H^n is Hausdorff n-dimensional surface measure. If, for example, S is composed of polygonal pieces S_i with oriented unit normals v_i, then $\Phi(S) = \sum_i \Phi(v_i) \operatorname{area}(S_i)$. Perhaps the most important integrands $\Phi : S^2 \to R$ arise as the surface free energy density functions for interfaces S between an ordered material A (hereafter called a crystal) and another phase or a crystal of another orientation. In this case $v_S(p)$ is the unit exterior normal to A at $p \in S$ and $\Phi(S)$ gives the surface free energy of S. Other interesting Φ's need not be continuous or even bounded. See the sailboat example of [T1], in which $\Phi(v)$ is the time required to sail unit distance in direction v rotated by 90°.

In this paper we survey what is known about the geometry of a single crystal A in equilibrium. In the special case in which A is a sessile or pendant crystal in a gravitational field and Φ is convex and invariant under all rotations about the vertical axis, we show (for the first time) that rotational symmetrization of A about the same axis does not increase total free energy.

We would like to acknowledge the partial support of both authors by NSF grants.

2. Examples of Integrands Φ

A few examples of Φ's are given, in order to illustrate some of the possibilities and to provide examples for the results to follow.

EXAMPLE 0. $\Phi_0(v) = 1$ for every v in S^n. Then $\Phi(S)$ is the area of S.

EXAMPLE 1. $\Phi_1(v) = |v_1| + |v_2| + |v_3|$ for $v = (v_1, v_2, v_3)$ in S^2.

EXAMPLE 2. $\Phi_2(v) = \max\{|v_1|, |v_2|, |v_3|\}$ for $v = (v_1, v_2, v_3)$ in S^2.

EXAMPLE 3. $\Phi_3(v) = (1 - v_3)^{1/2} + C|v_3|$ for $v = (v_1, v_2, v_3)$ in S^2.

EXAMPLE 4. $\Phi_4(v) = |v_1| + |v_2| + |v_3|$ for all $v = (v_1, v_2, v_3)$ in S^2 except $(\pm 1/\sqrt{3}, 1/\sqrt{3}, 1/\sqrt{3})$; for these v, $\Phi(v) = (5/6)\sqrt{3}$.

EXAMPLE 5. Let W be a compact convex body in R^{n+1}, and let Φ_W be the support function of W, restricted to the unit sphere. If the boundary of W is twice differentiable and has positive upper and lower bounds on its curvatures, the corresponding Φ_W is called an elliptic integrand.

3. Free Single Crystal Problem

Given Φ, what is the shape of an open region A of volume 1 which minimizes $\Phi(\partial A)$ among all regions with rectifiable boundaries having volume 1? There is a complete solution to this problem. The Wulff shape for Φ is defined to be

$$W_\Phi = \{x \in R^{n+1} : x \cdot v \leqslant \Phi(v) \text{ for all } v \text{ in } S^n\}.$$

Provided the interior of W_Φ is nonempty, the solution to this problem (unique up to translations) is the interior of W_Φ, scaled so that its volume is 1. If the interior of W_Φ is empty, there is no solution. See [T4] for a short clean proof of minimality, and see [T1] for references to other proofs.

The Wulff shapes for the examples above are as follows:

W_{Φ_0} is the unit ball $\{x: |x| \leqslant 1\}$ (which indeed is the shape of the region of least surface area compared to any other shape with the same volume).

W_{Φ_1} is the cube $\{x = (x_1, x_2, x_3): |x_i| \leqslant 1$ for $i = 1, 2, 3\}$.

W_{Φ_2} is the octahedron $\{x: x \cdot (\pm 1/\sqrt{3}, \pm 1/\sqrt{3}, \pm 1/\sqrt{3}) \leqslant 1\}$.

W_{Φ_3} is the right circular cylinder centered at the origin with axis in the x_3 direction, having radius 1 and height $2C$.

W_{Φ_4} is a cube with two of its eight corners truncated by triangular plane segments.

W_{Φ_W} is the W used to define Φ_W in Example 5.

One can extend Φ to a function on all of R^{n+1} by defining $\Phi(rv) = r\Phi(v)$ for any nonnegative r. Since any W_Φ is automatically convex and compact, the integrands of Example 5 in fact consist of all integrands which are convex (when so extended) and for which the free single crystal problem has a compact solution. However, not all nonelliptic integrands need be convex: Φ_4 is not convex, and whenever the collection of all oriented unit normals to W_Φ is not all of S^n there are nonconvex integrands Φ' for which $W_{\Phi'} = W_\Phi$ (each has $\Phi'(v) \geqslant \Phi(v)$, with equality on all v which are normals of W_Φ).

4. Crystals on a Substrate

This problem differs from the free single crystal problem in two ways:

(1) The crystal A is confined to a half space, which we take to be $x \cdot v < 0$ for some specified unit vector v (in the problem the orientations of A and v are fixed and are not allowed to rotate); $\{x : x \cdot v \leqslant 0\}$ is called the substrate.
(2) There is a specified value σ_v for the surface energy of the body in contact with the plane $x \cdot v = 0$ (more precisely, σ_v is the difference between the body-substrate surface energy per unit area and that of the substrate with the surrounding phase).

This problem is completely solved by the Winterbottom modification of the Wulff construction [W]:

$$W_{\Phi,v} = \{-\sigma_v v\} + W_\Phi \cap \{x \in R^{n+1} : x \cdot v \leqslant \sigma_v\} \quad \text{(set addition)}.$$

If $W_{\Phi,v}$ has a nonempty interior, then A has that shape, scaled so that it has the proper volume. If it has no interior, then there is no solution to the problem—the crystal spreads out to cover the entire plane. If $W_{\Phi,v}$ does not have v as a normal, then A need not touch the plane.

The question of what is the best orientation for the body relative to a given half space is the same as the question of which v gives the smallest energy for the properly scaled size of $W_{\Phi,v}$. As shown in [ATZ], the smallest energy occurs when $L^{n+1}(W_{\Phi,v})$ is smallest. (For $n = 2$, when A is $W_{\Phi,v}$ scaled to have volume V, its surface energy is $3V^{2/3}L^3(W_{\Phi,v})^{1/3}$.) Therefore, the answer depends quite strongly on σ_v and how it varies with v, as well as on Φ.

For $\Phi = \Phi_0$, the energy will be least when v is the direction where σ_v is the smallest (and there will be no solution if and only if $\sigma_v(v) \leqslant -1$ for some σ_v).

For $\Phi = \Phi_1$ and σ_v a constant, the best orientation depends strongly on that constant. If $\sigma_v > \sqrt{3}$, the crystal will not contact the table at all, and the best shape is that of W_{Φ_1}, regardless of v. For $\sigma_v = \sqrt{3} - \varepsilon$, the best orientation is $v = (1/\sqrt{3})(1, 1, 1)$ (and its seven other equivalent directions). For $\sigma_v = 0$, all orientations result in the same volume for $W_{\Phi_1,v}$ and thus all have equal surface energy. For σ_v slightly positive, the best orientation is $v = (1, 0, 0)$ (or one of its five equivalent directions). For σ_v slightly negative, $v = (1/\sqrt{3})(1, 1, 1)$ is again best since the cross section of W_{Φ_1} through the origin normal to that direction has the largest area. For $\sigma_v = -1$, the best orientation is $v = (1, 0, 0)$, and the crystal spreads out over the whole plane. For $\sigma_v < -1$, other directions achieve this degenerate minimum also, with all directions becoming minimizing at $\sigma_v \leqslant -\sqrt{3}$.

If neither Φ nor σ_v is a constant function, then very few general statements can be made. In any case, a slight perturbation of the example above serves to illustrate that the best orientation is not necessarily the one for which σ_v is the smallest.

It has recently been observed [ZAT] that a modification of Winterbottom's construction provides the solution for the shape of a crystal (of fixed orientation) in R^{n+1} in the angle of up to $n + 1$ n-dimensional planes (where $n = 2$ is the physically relevant case). The construction also works sometimes in the presence of yet more planes.

For more complicated fixed surfaces, there is no particular reason to believe that the surface of the crystal would be entirely convex or concave, and no general construction of solutions is known. However, the variety of possible local structures (tangent cones) composed of plane segments which cannot be infinitesimally corrugated without increasing energy has been catalogued in [TC1] and proved to be a complete catalog in [T3]. (A "second order," cusp-shaped singularity is also known to be an energy-minimizing shape for $\Phi = \Phi_3$ [TC2].)

5. Sessile and Pendant Crystals

The situation is considerably more complicated when there is a gravitational force acting. We will assume from now on that the substrate has normal $v = e_{n+1} = (0,\ldots,0,1)$ and that gravity acts in the direction $-e_{n+1}$ for sessile crystals (crystals sitting on a table) and in the direction e_{n+1} for pendant crystals. The problem is to minimize the total energy

$$\int_{p \in \partial A \cap \{x: x_{n+1} > 0\}} \Phi(v_{\partial A}(p))\, dH^n p + \sigma_v H^n(\partial A \cap \{x: x_{n+1} = 0\})$$

$$+ g\Delta\rho \int_{x \in A} x_{n+1}\, dL^{n+1}.$$

Here g is the gravitational constant and $\Delta\rho$ is the difference in density between the crystal and the surrounding phase. Provided a solution exists in the absence of gravity, a solution for the sessile crystal problem can be shown to exist (as a possibly "infinitesimally corrugated" varifold if Φ is not convex) by general compactness arguments of geometric measure theory and by using the technique of [A1 VI.2(1)] to keep a minimizing sequence from wandering off to infinity. For a pendant crystal, existence of a solution presumably depends on limiting the volume (or $g\Delta\rho$) and restricting the body to lie below another parallel plane. Uniqueness of solutions is not known for either problem. Furthermore, it is not known that a sessile crystal must be a convex body (even though it is always convex in the absence of gravity). Finally, if $W_{\Phi, -e_{n+1}}$ is rotationally invariant about e_{n+1}, one might expect that *any* solution to the problem would also be symmetric in the presence of gravity; this is also not known to be true. In Sections 6–10 of this paper we do, however, present the first proof that *some* solution has this symmetry.

If a solution A to the sessile crystal problem is convex, then various additional properties of possible crystal faceting can be established; the following theorem sets forth most of this additional knowledge.

Theorem [ATZ] [T2]. *Let A be an $(n + 1)$-dimensional region of volume 1 contained in the upper half space and having least total energy (surface energy plus gravitational energy) in a gravitational field. If $g\Delta\rho$ is large enough, if A is convex (an open problem!), and if all unit vectors in a neighborhood of the vertical vector e_{n+1} do not occur as normals to W_Φ (so that W_Φ has a corner on top), then there is a facet on top of A with vertical normal. If $n = 2$ and W has an edge on top (perpendicular to e_{n+1}), then under some conditions there must be a facet on top of A and under others there can be no such facet but there must be gravity-induced curvature near the top of A. In either case, if Φ is not a convex integrand in the region around the vertical unit vector, this facet or curved region must be infinitesimally corrugated—that is, A is not a classical solution but only a varifold solution.*

6. General Terminology

For the rest of this paper, we fix $n \in \{2, 3, 4, \ldots\}$ and use the following assumptions and terminology.

(1) $\Phi: R^{n+1} \to R^+$ is a parametric integrand (i.e., an integrand on S^n extended as in Section 3 to all of R^{n+1}) which is *convex* and *invariant* under all rotations about the x_{n+1} axis. (For nonconvex integrands Φ', one replaces Φ' by the convex integrand Φ with $W_\Phi = W_{\Phi'}$, proves the theorems with Φ, and then observes that any surface can be "infinitesimally corrugated" so that its Φ' integral is the same as its Φ integral [T1 3.3].) We define $F: R^2 \to R^+$ by requiring $F(u, v) = \Phi(u, 0, \ldots, 0, v)$ for each u, v. We note that F is also convex and positively homogeneous of degree 1 with $F(u, v) = F(-u, v)$ for each u, v.

(2) If A is an L^{n+1} measurable subset of R^{n+1} we denote by A_* the associated set consisting of points p in R^{n+1} for which $\Theta^{n+1}(L^{n+1} \sqcup A, p) = 1$. It is well known that $A = A_*$ except possibly for a set of L^{n+1} measure 0. Each L^{n+1} measurable subset A of R^{n+1} has a well defined measure-theoretic boundary ∂A (consisting of those points in R^{n+1} at which neither A nor its complement $R^{n+1} \sim A$ has density one), and A is said to have finite perimeter provided $H^n(\partial A) < \infty$. If A does have finite perimeter then the set of points p in R^{n+1} at which A has a measure theoretic unit exterior normal vector $n(A, p)$ is denoted $\partial^* A$. It is well known in such a case that $\partial A = \partial^* A$ except possibly for a set of H^n measure 0. Corresponding terminology applies if B is an H^n measurable subset of some n-dimensional affine subspace of R^{n+1}.

(3) Suppose A is a bounded L^{n+1} measurable subset of R^{n+1}. For each $t \in R$ we denote by $r(t)$ that nonnegative radius for which $H^n[U^n(0, r(t))] = H^n[A \cap \{x : x_{n+1} = t\}]$; here $U^n(0, r) = R^n \cap \{x : |x| < r\}$. Associated with

A is then a symmetrized set

$$\mathrm{SYM}(A) = \left[\bigcup_{t \in R} U^n(0, r(t)) \times \{t\} \right]_*.$$

It is essentially a consequence of Fubini's theorem that the union of these stacked disks is L^{n+1} measurable and has the same measure as A. "$*$"-ing does not change this measure nor the measure of almost every cross section; it, however, in many cases gives an intuitively more natural symmetrized solid.

7. Heuristic Description of the Basic Symmetrization Inequalities

Suppose $A \subset R^{n+1}$, $-\infty < t < \infty$, $B = A \cap R^n \times \{t\}$ is the cross section of A at height t, $B_0 = \mathrm{SYM}(A) \cap R^n \times \{t\}$ is the corresponding cross section of $\mathrm{SYM}(A)$ at height t, $C = \partial B$, $C_0 = \partial B_0$; note [F 4.3.1, 4.3.8]. Suppose also that $\theta_* \in (0, \pi)$ is the "cotangent average" angle that the exterior normals $n(A, p)$ make with e_{n+1} for $p \in C$. Similarly, let $\theta_0 \in (0, \pi)$ denote the angle that the exterior normals $n(\mathrm{SYM}(A), p)$ make with e_{n+1} for $p \in C_0$. Then:

(1) The facts that $H^n(B) = H^n(B_0)$ and B_0 is a disk (by our symmetrization construction) imply that $H^{n-1}(C_0) \leqslant H^{n-1}(C)$—the usual "isoperimetric inequality."
(2) The fact that A and $\mathrm{SYM}(A)$ have the same cross sectional areas at each height implies that the rates of changes of these cross sectional areas as a function of height are also the same. For A this rate of change equals $-H^{n-1}(C)\cot(\theta_*)$ at height t, while for $\mathrm{SYM}(A)$ this rate is $-H^{n-1}(C_0)\cot(\theta_0)$. Hence, $H^{n-1}(C)\cot(\theta_*) = H^{n-1}(C_0)\cot(\theta_0)$.
(3) The rate of increase of Φ integral over $\partial^* A \cap R^n \times (-\infty, t)$ with height t equals

$$H^{n-1}(C_*)F(\sin(\theta_*), \cos(\theta_*))\csc(\theta_*) = H^{n-1}(C)F(1, \cot(\theta)),$$

while the corresponding rate of increase of Φ integral over $\partial^* \mathrm{SYM}(A) \cap R^n \times (-\infty, t)$ equals $H^{n-1}(C_0)F(1, \cot(\theta_0))$. Using (2) we set

$$0 < \rho = H^{n-1}(C_0)/H^{n-1}(C) = \cot(\theta_*)/\cot(\theta_0) \leqslant 1$$

and estimate

$H^{n-1}(C)F(1, \cot(\theta_*))$

$\qquad \geqslant H^{n-1}(C)F(\rho, \cot(\theta_*))$ (since F is convex and $F(u, v) = F(-u, v)$)

$\qquad = H^{n-1}(C)\rho F(1, \cot(\theta_*)/\rho)$ (since F is positively homogeneous of degree 1)

$\qquad = H^{n-1}(C_0)F(1, \cot(\theta_0))$.

This gives the basic symmetrization inequality that the Φ integral of $\partial^* \mathrm{SYM}(A)$ does not exceed the Φ integral of $\partial^* A$.

Since A is a general measurable set having a rectifiable set as perimeter we, of course, must implement these heuristic ideas and inequalities in a measure theoretic way. The basic tools are Fubini's theorem, Federer's coarea formulas, and the Gauss–Green theorem. The heuristic arguments above do not, of course, apply at heights t for which $H^n(\partial^* A \cap R^n \times \{t\}) > 0$.

8. Consequences of Federer's Coarea Formulas and the Gauss–Green Theorem

Suppose:

(a) A is a bounded L^{n+1} measurable subset of R^{n+1} with $A = A_*$ and $H^n(\partial A) < \infty$.

(b) $\theta: \partial^* A \to [0, \pi]$ is given by requiring $\cos[\theta(p)] = e_{n+1} \cdot n(A, p)$ for $p \in \partial^* A$, i.e., $\theta(p)$ is the angle between e_{n+1} and $n(A, p)$.

(c) $\pi: \partial^* A \to R$ is given by setting $\pi(x) = x_{n+1}$ for $x \in \partial^* A$.

(d) For $s \in R$ we set

$$A(s) = A \cap \{x: x_{n+1} < s\}, \quad \partial^* A(s) = (\partial^* A) \cap \{x: x_{n+1} < s\},$$

$$B(s) = A \cap \{x: x_{n+1} = s\}, \quad C(s) = (\partial^* A) \cap \{x: x_{n+1} = s\}.$$

Then

(1) For H^n almost every $x \in \partial^* A$, the approximate one-dimensional Jacobian ap $J_1 \pi(x)$ equals $\sin[\theta(x)]$.

(2) For $L^1 \times L^1$ almost every (s, t) with $-\infty < s < t < \infty$, the positive homogeneity of F combined with (1) and the coarea formula [F 3.2.22] give

$$\int_{x \in \partial^* A(t) \sim \partial^* A(s)} \Phi(n(A, x)) \, dH^n x$$

$$= \int_{x \in \partial^* A(t) \sim \partial^* A(s)} F(\sin[\theta(x)], \cos[\theta(x)])(\sin[\theta(x)])^{-1} \text{ ap } J_1 \pi(x) \, dH^n x$$

$$= \int_{x \in \partial^* A(t) \sim \partial^* A(s)} F(1, \cot[\theta(x)]) \text{ ap } J_1 \pi(x) \, dH^n x$$

$$= \int_{r \in (s, t)} \int_{x \in C(r)} F(1, \cot[\theta(x)]) \, dH^{n-1} x \, dL^1 r$$

and

$$\int_{x \in \partial^* A(t) \sim \partial^* A(s)} x_{n+1} \cos[\theta(x)] \, dH^n x$$

$$= \int_{x \in \partial^* A(t) \sim \partial^* A(s)} x_{n+1} \cot[\theta(x)] \text{ ap } J_1 \pi(x) \, dH^n x$$

$$= \int_{r \in (s, t)} \int_{x \in C(r)} x_{n+1} \cot[\theta(x)] \, dH^{n-1} x \, dL^1 r.$$

(3) The function

$$s \to \int_{x \in \partial^* A(s)} \Phi(n(A, x)) \, dH^n x$$

is differentiable at L^1 almost every s with derivative equal to

$$\int_{x \in C(s)} F(1, \cot[\theta(x)]) \, dH^{n-1} x.$$

(4) The function

$$s \to \int_{x \in \partial^* A(s)} x_{n+1} \cos[\theta(x)] \, dH^n x$$

is differentiable at L^1 almost every s with derivative equal to

$$\int_{x \in C(s)} s \cot[\theta(x)] \, dH^{n-1} x.$$

(5) It follows from Fubini's theorem that the function $s \to L^{n+1}(A(s))$ is absolutely continuous with derivative equal to $H^n(B(s))$ at L^1 almost every s.

(6) For $L^1 \times L^1$ almost every $(t, t + \Delta t)$ with $-\infty < t < t + \Delta t < \infty$ we can then apply the Gauss–Green theorem with the vectorfield $v: R^{n+1} \to R^{n+1}$, $v(x) = x_{n+1} e_{n+1}$ for each x, to obtain

$$L^{n+1}[A(t + \Delta t) \sim A(t)]$$

$$= \int_{A(t+\Delta t) \sim A(t)} \operatorname{div}(v) \, dL^{n+1}$$

$$= \int_{x \in \partial^*[A(t+\Delta t) \sim A(t)]} v(x) \cdot n(A(t + \Delta t) \sim A(t), x) \, dH^n x$$

$$= (t + \Delta t) H^n(B(t + \Delta t)) - t H^n(B(t))$$

$$\quad + \int_{x \in \partial^* A(t+\Delta t) \sim \partial^* A(t)} x_{n+1} \cos[\theta(x)] \, dH^n x$$

$$= t[H^n(B(t + \Delta t)) - H^n(B(t))] + \Delta t H^n(B(t + \Delta t))$$

$$\quad + \int_{x \in \partial^* A(t+\Delta t) \sim \partial^* A(t)} x_{n+1} \cos[\theta(x)] \, dH^n x.$$

(7) If the functions of (4) and (5) are differentiable at $s = t$ and, additionally, the function $s \to H^n(B(s))$ is continuous at $s = t$, then this last function is differentiable at $s = t$ with

$$-(d/ds) H^n(B(s))|_{s=t} = \int_{x \in C(t)} \cot[\theta(x)] \, dH^{n-1} x.$$

9. Proposition

Hypotheses:

(a) A, θ, etc. are as in (a), (b), (c), (d) of Sec. 8 above, and $A_0 = \mathrm{SYM}(A)$.
(b) $H^n(\partial A_0) < \infty$ and $\theta_0 \colon \partial^* A_0 \to [0, \pi]$, $A_0(s)$, $\partial^* A_0(s)$, $B_0(s)$, $C_0(s)$ have the obvious meanings associated with Sec. 8(b)(c)(d), with A_0 replacing A there.
(c) $-\infty < r < t < \infty$.
(d) $H^n(\partial^* A \cap \{x \colon r \leqslant x_{n+1} \leqslant t \text{ and } \theta(x) = 0\}) = 0$.
(e) For $r \leqslant s \leqslant t$ the functions of Sec. 8(3) and (4) are absolutely continuous in s.
(f) For $r \leqslant s \leqslant t$ the function $s \to H^n(B(s))$ is continuous in s.
(g) The conditions of (d), (e), (f) above also hold, with A replaced by A_0, etc.

Conclusion:

$$\int_{p \in \partial^* \mathrm{SYM}(A) \cap \{x \colon r < x_{n+1} < t\}} \Phi(n(\mathrm{SYM}(A), p)) \, dH^n p$$

$$\leqslant \int_{p \in \partial^* A \{x \colon r < x_{n+1} < t\}} \Phi(n(A, p)) \, dH^n p.$$

PROOF. The proof is in several steps.

Step 1. It follows from General Terminology 6(3) and Hypotheses (f), (g) above that $H^n(B(s)) = H^n(B_0(s))$ for each $r \leqslant s \leqslant t$.
Step 2. It follows from Conclusion (1) above together with 8(7) above (applied to both $B(s)$ and $B_0(s)$) that, for L^1 almost every s with $r \leqslant s \leqslant t$,

$$\int_{C_0(s)} \cot(\theta_0) \, dH^{n-1} = \int_{C(s)} \cot(\theta) \, dH^{n-1}.$$

Step 3. It follows from 6(3) and the well known extremal properties of spheres that, for L^1 almost every s with $r \leqslant s \leqslant t$, $H^{n-1}(C_0(s)) \leqslant H^{n-1}(C(s))$.
Step 4. It follows from our condition that $H^n(\partial^* A) < \infty$ (part of Hypothesis (a)) that for almost every s with $r \leqslant s \leqslant t$ either $H^{n-1}(C(s)) = 0$ or there exists $\theta_*(s) \in (0, \pi)$ with

$$H^{n-1}(C(s)) \cot[\theta_*(s)] = \int_{C(s)} \cot(\theta) \, dH^{n-1}.$$

For those s for which $\theta_*(s)$ is so defined, the convexity of the function $v \to F(1, v)$ then implies that

$$H^{n-1}(C(s)) F(1, \cot[\theta_*(s)]) \leqslant \int_{C(s)} F(1, \cot(\theta)) \, dH^{n-1}.$$

For almost every s with $r < s < t$ either $H^{n-1}(C_0(s)) = 0$ or $\theta_0 | C_0(s)$

is constant—we denote this constant value by $\theta_0(s)$—so that

$$H^{n-1}(C_0(s))F(1,\cot[\theta_0(s)]) = \int_{x \in C_0(s)} F(1,\cot(\theta_0(x)))\,dH^{n-1}x.$$

Step 5. For those s with $H^{n-1}(C(s)) > 0$ and for which the definitions and conclusions of Steps 2, 3, and 4 apply, we define $0 < \rho \leqslant 1$ by setting $\rho = H^{n-1}(C_0(s))/H^{n-1}(C(s))$ and infer from Steps 2 and 4 that

$$\rho H^{n-1}(C(s))\cot(\theta_0(s)) = H^{n-1}(C_0(s))\cot(\theta_0(s))$$
$$= H^{n-1}(C(s))\cot(\theta_*(s)).$$

Now either $\cot[\theta_1(s)] = \cot[\theta_*(s)] = 0$ or $\cot[\theta_1(s)] \neq 0 \neq \cot[\theta_*(s)]$ and $\rho = \cot[\theta_*(s)]/\cot[\theta_1(s)]$; in the first case the inequality

$$H^{n-1}(C_0(S))F(1,\cot[\theta_0(s)]) \leqslant H^{n-1}(C(s))F(1,\cot[\theta_*(s)])$$

is clear while in the second case this inequality follows from the estimates of the heuristic argument of 7(3) (the foundations of which have now been made rigorous). Using Step 4, we conclude that

$$\int_{x \in C_0(s)} F(1,\cot[\theta_0(x)])\,dH^{n-1}x \leqslant \int_{x \in C(s)} F(1,\cot[\theta(x)])\,dH^{n-1}x$$

for almost all s with $r \leqslant s \leqslant t$, which in view of 8(3) and Hypotheses (e) and (g) above is sufficient to prove the conclusion of the present proposition.   □

10. Theorem (Symmetrization Does Not Increase Convex Parametric Integrals)

Suppose A is a bounded L^{n+1} measurable subset of R^{n+1} with $H^n(\partial A) < \infty$. Suppose also $\Phi: R^{n+1} \to R^+$ is a convex parametric integrand which is invariant under all rotations about the x_{n+1} axis. Then

$$\int_{p \in \partial^* \mathrm{SYM}(A)} \Phi(n(\mathrm{SYM}(A),p))\,dH^n p \leqslant \int_{p \in \partial^* A} \Phi(n(A,p))\,dH^n p.$$

Furthermore,

$$\int_{x \in \mathrm{SYM}(A)} f(x_{n+1})\,dL^{n+1}x = \int_{x \in A} f(x_{n+1})\,dL^{n+1}x$$

for every L^1 summable function $f: R \to R$; in particular gravitational energy is not altered.

PROOF. We assume without loss of generality that $A = A_*$. The first assertion follows from Proposition 9 in case the hypotheses there hold for each $-\infty < r < t < \infty$. In the general case one carefully approximates A in L^{n+1} measure by polyhedral solids P in general position to which such hypotheses apply and for which $H^n(\partial P)$ nearly equals $H^n(\partial^* A)$—this implies that the corresponding Φ integrals are also close. Such solids P are readily obtained, for example, by use in part of the Approximation Theorem [F 4.2.20] together with a slight Euclidean motion. The lower semicontinuity of convex parametric integrals under flat convergence implies the domination of the Φ integral over $\partial^* \text{SYM}(A)$ by the Φ integral over $\partial^* A$ as required. The second assertion follows from the definition of $\text{SYM}(A)$. $\qquad\square$

Caution. Various examples show that a condition such as $H^n(\partial^* \text{SYM}(A)) = H^n(\partial^* A)$ does not imply that A_* is already rotationally symmetric, even if the center of gravity of A is above the origin; typically such examples will contain pieces of $R^n \times \{t\}$ for one or more t's.

11. Remark.

The symmetrization technique discussed above is similar to the spherical symmetrization technique set forth in the companion paper [A2]. We also call attention to the multiple valued function averaging technique set forth in [AS] which similarly (when applicable) preserves volumes while not increasing convex surface integrals.

References

[A1] F.J. Almgren, Jr., *Existence and regularity almost everywhere of solutions to elliptic variational problems with constraints*, Mem. Amer. Math. Soc. **165** (1976).

[A2] —, *Spherical symmetrization*, Integral Functionals in the Calculus of Variations, Lecture Notes in Mathematics, Springer-Verlag, New York, 1986.

[AS] — and B. Super, *Multiple valued functions in the geometric calculus of variations*, Astérisque **118** (1984), 13–22.

[ATZ] J.E. Avron, J.E. Taylor, and R.K.P. Zia, *Equilibrium shapes of crystals in a gravitational field: Crystals on a table*, J. Stat. Phys. **33** (1983), 493–522.

[F] H. Federer, *Geometric Measure Theory*, Springer-Verlag, New York, 1969.

[T1] J.E. Taylor, *Crystalline variational problems*, Bull. Amer. Math Soc. **84** (1978), 568–588.

[T2] —, *Is there gravity induced faceting in crystals?* Astérisque **118** (1984), 243–253.

[T3] —, *Complete catalog of minimizing embedded crystalline cones*, Proc. Symp. Pure Math. **44** (1985).

[T4] —, *Some crystalline variational techniques and results*, Minimal Surface Seminar, Ecole Polytechnique (to appear in Astérisque).

[TC1] J.E. Taylor and J.W. Cahn, *Catalog of saddle shaped surfaces in crystals*, Acta Metallurgica **34** (1986), 1–12.

[TC2] —, *A cusp singularity in surfaces that minimize an anisotropic surface energy*, Science **233** (1986), 548–551.

[W] W.L. Winterbottom, *Equilibrium shape of a small particle in contact with a foreign substrate*, Acta Metallurgica **15** (1967), 303–310.

[ZAT] R.K.P. Zia, J.E. Avron, and J.E. Taylor, *Crystals in corners: the summertop construction* (in preparation).

Immersed Tori of Constant Mean Curvature in R^3

Henry C. Wente

In this chapter we show how to construct immersions of tori into Euclidean space R^3 which have constant mean curvature $H \neq 0$. We thus exhibit an example of a "non-round" soap bubble (although it does self-intersect) providing a counterexample to a conjecture attributed to H. Hopf. We shall carefully state the theorems involved in the construction and also provide a geometric description (with suggestive sketches) of the desired surfaces. An expanded version complete with proofs appeared in a recent paper of the author [11].

Conjecture of H. Hopf. *If Σ is an immersion of an oriented closed hypersurface in R^n with constant mean curvature $H \neq 0$, then the hypersurface Σ is the round embedded $(n-1)$-sphere.*

An early result relating to this problem was made back in 1853 by J.H. Jellett [7]. He showed that if Σ is a star shaped surface in R^3 (i.e. may be represented by a radial graph over a sphere) with constant mean curvature, then Σ is the round sphere. This result was improved by A.D. Alexandroff [2] who showed that if Σ is an embedded closed hypersurface of constant mean curvature $H \neq 0$ in R^n of any genus, then Σ must be the round $(n-1)$-sphere. H. Hopf [5] proved that if Σ is an immersion of a sphere S^2 into R^3 with constant mean curvature then Σ is just the round sphere. Recently, Wu-Yi Hsiang [6] produced an immersion of S^3 into R^4 with constant mean curvature $H \neq 0$ which was not the round sphere, thus providing a counterexample to Hopf's conjecture in higher dimensions. This example put to rest the possibility that the Alexandroff technique might somehow apply to immersed surfaces which were not embeddings. However, Hsiang's construction does not extend to the classical dimension ($= 3$), and the conjecture has remained unsettled in this case. We have the following result.

Counterexample Theorem. *There exist closed immersed surfaces of genus one in R^3 with constant mean curvature. (In fact, we exhibit a countably infinite number of isometrically distinct examples.)*

We shall exhibit the surface by producing a conformal mapping of the plane R^2 into R^3 with constant mean curvature which is doubly periodic with respect to a rectangle in the plane. Let $w = (u, v) = u + iv$ represent a typical point in $R^2 = C$ while $\bar{x} = (x, y, z)$ denotes a point in R^3 so that our immersion is given by a function $\bar{x}(u, v)$. We let

$$d\bar{x} \cdot d\bar{x} = ds^2 = E(du^2 + dv^2) = e^{2\omega}(du^2 + dv^2) \tag{1a}$$

$$-d\bar{x} \cdot d\bar{\xi} = L\,du^2 + 2M\,du\,dv + N\,dv^2 \tag{1b}$$

be the first and second fundamental forms for the surface. We shall set the mean curvature $H = \frac{1}{2}$. The Gauss and Codazzi–Mainardi equations in this case become (see [5] for details)

$$\Delta\omega + Ke^{2\omega} = 0, \quad K = \text{Gauss curvature} = (LN - M^2)/E^2 \tag{2a}$$

$$\phi(w) = [(L - N)/2] - iM \text{ is a complex analytic function.} \tag{2b}$$

If $\bar{x}(u, v)$ is a doubly periodic map then so are the functions $\omega(u, v)$ and $\phi(w)$ listed above. Since $\phi(w)$ is also complex analytic it must be a constant. Now

$$|\phi(w)| = |k_1 - k_2|E/2,$$

where k_1, k_2 are the principle curvatures of the surface. If $\phi(w) \equiv 0$, then $k_1 = k_2$ and every point on the immersed surface would be an umbilic point. This is impossible. Thus $\phi(w) = \text{constant} \neq 0$, and the immersed surface is free of umbilics.

The preimage of the lines of curvature in the (u, v)-plane are determined by the form

$$-M\,du^2 + (L - N)\,du\,dv + M\,dv^2 = 0.$$

We observe that this form is a constant in the plane showing that the lines of curvature correspond to a family of orthogonal straight lines. We present our construction so that these lines are parallel to the coordinate axes, which is the case when $M = 0$.

Now suppose that $\omega(u, v)$ is a solution to the differential equation

$$\Delta\omega + \sinh\omega \cosh\omega = 0. \tag{3}$$

If we set $E = e^{2\omega}$, $L = e^{\omega}\sinh\omega$, $M = 0$, and $N = e^{\omega}\cosh\omega$, then it follows that the Gauss and Codazzi–Mainardi equations are satisfied, and by a theorem of Bonnet the system can be integrated to yield a surface $\bar{x}(u, v)$, unique up to a Euclidean motion in R^3, having the given fundamental forms.

The equations to be integrated are

$$\bar{x}_{uu} = \omega_u \bar{x}_u - \omega_v \bar{x}_v + L\bar{\xi}$$
$$\bar{x}_{uv} = \omega_v \bar{x}_u + \omega_u \bar{x}_v + M\bar{\xi}$$
$$\bar{x}_{vv} = -\omega_u \bar{x}_u + \omega_v \bar{x}_v + N\bar{\xi} \qquad (4)$$
$$\bar{\xi}_u = -k_1 \bar{x}_u$$
$$\bar{\xi}_v = -k_2 \bar{x}_v.$$

Here $k_1 = L/E = e^{-\omega}\sinh\omega$, $k_2 = e^{-\omega}\cosh\omega$, so we see that the lines of curvature correspond to lines parallel to the coordinate axes in R^2. Furthermore, the surface is free of umbilic points and has constant mean curvature $H = \frac{1}{2}$.

If $\bar{x}(u, v)$ is to be a doubly periodic mapping, then so must $\omega(u, v)$. However, the converse need not be true. Suppose that $\omega(u, v)$ is a positive solution to the differential equation (3) on a rectangular domain Ω_{AB} lying in the first quadrant with two of its sides on the coordinate axes and the vertex opposite the origin at (A, B). Suppose also that the solution $\omega(u, v)$ vanishes on the boundary of the rectangle. Following the argument used in [4], one can show that $\omega(u, v)$ satisfies the following symmetry properties.

(a) $\omega(u, v)$ is symmetric about the lines $u = A/2$ and $v = B/2$.
(b) For a fixed $v, 0 < v < B$, $\omega(u, v)$ is an increasing function of $u, 0 \leqslant u \leqslant A/2$. For a fixed $u, 0 < u < A$, $\omega(u, v)$ is an increasing function of v, $0 \leqslant v \leqslant B/2$. $\qquad (5)$
(c) $\omega_v(u, 0)$ is strictly increasing for $0 \leqslant u \leqslant A/2$.
$\omega_u(0, v)$ is strictly increasing for $0 \leqslant v \leqslant B/2$.

Furthermore, $\omega(u, v)$ can be extended as a solution of the differential equation (3) on all of R^2 by odd reflections across the grid lines $u = mA, v = nB$ (m, n integers).

Theorem 2. *Suppose $\omega(u, v)$ is a solution to the differential equation (3) on R^2 which is positive on the fundamental rectangle Ω_{AB}, vanishing on the boundary, and satisfying the properties (5). The mapping $\bar{x}(u, v)$ obtained by integrating the system (4) is an immersed surface of constant mean curvature $H = 1/2$ and satisfying the following symmetry properties:*

(a) *The curve $\bar{x}((m + 1/2)A, v)$ lies in a normal plane Π_m with $\bar{x}_u$ as a normal vector to Π_m. If $\mathcal{R}_m$ is the reflection map about Π_m in R^3, then $\bar{x}((m + 1/2)A + u, v) = \mathcal{R}_m \circ \bar{x}((m + 1/2)A - u, v)$.*
(b) *The curve $\bar{x}(u, (n + 1/2)B)$ lies in a normal plane Ω_n with $\bar{x}_v$ as a normal vector to Ω_n. If $\mathcal{S}_n$ is the reflection map about Ω_n in R^3, then $\bar{x}(u, (n + 1/2)B + v) = \mathcal{S}_n \circ \bar{x}(u, (n + 1/2)B - v)$. Each Ω_n is orthogonal to each Π_m.*
(c) *The curve $\bar{x}(u, 0)$ is a planar curve lying in a plane Γ_0 which is a tangent plane to the surface at each point. This curve intersects each plane Π_m*

orthogonally. $\bar{x}_u(u, 0)$ is an even function of u. This allows us to conclude that all of the planes Π_m are parallel.

(d) The curve $\bar{x}(0, v)$ satisfies the condition $(\bar{x} + \bar{\xi})(0, v) = \bar{c}_0$ a constant vector. Therefore $\bar{x}(0, v)$ lies on a sphere $S(\bar{c}_0, 1)$ with center $\bar{c}_0$ and radius one. Similarly $\bar{x}(kA, v)$ lies on a sphere $S(\bar{c}_k, 1)$. The points $\bar{c}_k$ lie in every plane Ω_n.

(e) $\bar{x}(u + 2A, v) = \bar{x}(u, v) + \bar{b}$ where $\bar{b} = \bar{c}_2 - \bar{c}_0$ is a vector normal to the planes Π_m carrying Π_0 to Π_2.

(f) $\bar{x}(u, v + 2B) = \Theta \circ \bar{x}(u, v)$ where Θ is a rotation from Ω_0 to Ω_2 about their line of intersection 1. $\qquad(6)$

The surface will close up if we can select the rectangle Ω_{AB} so that the translation $\bar{b} = 0$ (i.e. all the planes Π_m are identical) and so that the rotation angle θ is a rational multiple of 2π. We use a continuity argument to show that this is possible. The procedure is as follows. Map (via a homothety) all rectangles of similar shape onto a representative rectangle which we select by the standard Schwartz–Christoffel mapping of rectangles onto the unit disk.

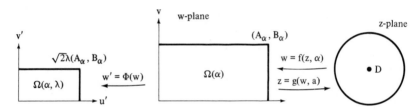

FIGURE 1. The fundamental domain.

We have the following identities satisfied by the various functions defined on the domains pictured in Figure 1.

(a) On $\Omega(\alpha, \lambda)$.

$\Delta\omega + \sinh\omega\cosh\omega = 0$.

$\Delta\sigma + \sinh\sigma = 0$, where $\sigma = 2\omega$.

(b) On $\Omega(\alpha)$.

$\Delta W + 2\lambda\sinh W\cosh W = 0$, where $W = \omega \circ \Phi$. $\qquad(7)$

$\Delta\Sigma + 2\lambda\sinh\Sigma = 0$, where $\Sigma = 2W$.

(c) On the disk D.

$\Delta\Psi + \lambda|f'(z, \alpha)|^2(e^\Psi - e^{-\Psi}) = 0$, where $\Psi = \Sigma \circ f$:

$$w = f(z, \alpha) = \int_0^z (t^4 + 2(\cos 2\alpha)t^2 + 1)^{-1/2}\, dt.$$

The proof of the existence of positive solutions to the system (7c) on D which vanish on the boundary is based on a method developed by V.K. Weston [10] and R.L. Moseley [9]. A key feature is that for small positive λ one obtains large solutions as indicated in the following theorem.

Theorem 3. *There exists an open set* $0 \subset (\alpha, \lambda)$*-plane where for each* α_1, α_2 *with* $0 < \alpha_1 < \alpha_2 < \pi/2$ *there exists* $\tilde{\lambda} = \tilde{\lambda}(\alpha_1, \alpha_2) > 0$ *so that* $[\alpha_1, \alpha_2] \times (0, \tilde{\lambda}] \subset 0$, *and a mapping from* 0 *to* $C(D)$ *denoted by* $\Psi(z, \alpha, \lambda)$ *such that*

(a) $\Sigma(w, \alpha, \lambda) = \Psi(g(w, \alpha), \alpha, \lambda)$ *is a positive solution to* (7b) *which vanishes on the boundary.*

(b) *The functions* $\Sigma, \Sigma_u, \Sigma_v$ *depend continuously on* (α, λ) *down to* $\lambda = 0$, *with* $\Sigma(w, \alpha, 0) = \Sigma_0(w, \alpha) = 4\log(1/|g(w, \alpha)|)$.

(c) *For* $\lambda > 0$ *the mapping* $(\alpha, \lambda) \to \Psi(z, \alpha, \lambda)$ *is a continuously differentiable mapping of* 0 *into* $C(D)$.

Remark on the Proof. One first constructs a good approximate solution $U_0(z, \lambda)$ with the correct asymptotic limit as λ approaches 0 by using the Liouville form of the exact solution to the differential equation $\Delta V + \lambda e^V = 0$, namely $\lambda e^V = |F'(z)|^2/(1 + |F(z)|^2)^2$, where $F(z)$ is a complex analytic function with at most simple zeros and poles. Then one applies a modified Newton iteration scheme, starting with $U_0(z, \lambda)$ using the appropriate integral operator, and shows that the resulting sequence converges in $C(\bar{D})$ to the desired solution.

We continue with our construction of the counterexample. We want to measure the distance between the parallel planes Π_0 and Π_1 and wish to show that for certain (α, λ) the distance is zero. It is better to look at the surfaces $\bar{y}(w, \alpha, \lambda) = \bar{x} \circ \Phi(w, \alpha, \lambda)/\sqrt{2\lambda}$ defined relative to the fundamental domain $\Omega(\alpha)$ and to measure the distance between the parallel planes Π_0 and Π_1 which correspond to the mapping $\bar{y}$. We do this by looking at the curve $\bar{y}(u, 0, \alpha, \lambda)$, a planar curve which cuts through the planes Π_m orthogonally and has the symmetry indicated in Figure 2.

The functions $\bar{y}(u, v, \alpha, \lambda)$ are conformal immersions into R^3 with constant mean curvature $H = \sqrt{2\lambda}$, so that as λ approaches 0 the mean curvature approaches 0 and the mapping tends to a planar map. The functions

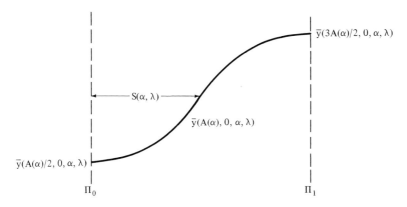

FIGURE 2. Measuring the distance between the parallel planes Π_0 and Π_1.

$\bar{y}$ satisfy a system just like (4) with ω replaced by $W = \Sigma/2$, L is replaced by $\tilde{L} = \sqrt{2\lambda L}$, and so on. Since by Theorem 3b the function $W(u, v, \alpha, \lambda)$ approaches $W(u, v, \alpha, 0) = 2\log(1/|g(w, \alpha)|)$ as λ approaches 0, the curve $\bar{y}(u, 0, \alpha, \lambda)$ approaches a limit curve $\bar{y}(u, 0, \alpha, 0)$ as λ approaches 0. It follows that the distance function $S = S(\alpha, \lambda)$, as indicated in Figure 2, is continuous down to $\lambda = 0$ and differentiable if λ is positive. Since $W(u, v, \alpha, 0)$ is known explicitly, one can calculate $S(\alpha, 0)$, obtaining

$$S(\alpha, 0) = \int_0^\beta (\cos 2\theta/(2\cos 2\theta - 2\cos 2\beta)^{1/2})\, d\theta, \qquad \beta = (\pi/2) - \alpha. \qquad (8)$$

We immediately have the following conclusions:

(a) $S(\alpha, 0)$ is strictly increasing for $0 < \alpha < \pi/2$;
(b) $S(\alpha, 0)$ approaches $-\infty$ as α approaches 0;
(c) $S(\alpha, 0)$ is positive for α greater than $\pi/4$.

It follows that there is exactly one value $\alpha^*, 0 < \alpha^* < \pi/4$, for which $S(\alpha^*, 0) = 0$. We have the following picture (see Figure 3). There is a small rectangle $[\alpha_1, \alpha_2] \times [0, \tilde{\lambda}]$ with $S(\alpha_1, \lambda)$ negative, $S(\alpha_2, \lambda)$ positive, and $S(\alpha^*, 0) = 0$. There is a connected set X included in this small rectangle on which S vanishes and which separates the left side of the rectangle from the right side. In particular, $(\alpha^*, 0)$ is in the set X and every line $\lambda = $ constant slices into X.

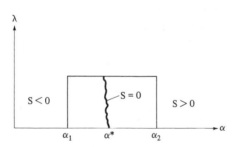

FIGURE 3. The set $S = 0$: all the planes Π_m are identical.

Now we measure the rotation angle between the planes Ω_0 and Ω_1 by looking at the image of the curve $\bar{y}(0, v, \alpha, \lambda)$, $B(\alpha)/2 < v < 3B(\alpha)/2$. From Theorem 2(d) it follows that this curve lies on the sphere with center $\bar{c}_0$ and radius $(2\lambda)^{-1/2}$, connecting the planes Ω_0 to Ω_1 and intersecting them orthogonally. Let $T(\alpha, \lambda)$ be the distance between these planes as measured on a great circle of the sphere whose radius is $(2\lambda)^{-1/2}$. By repeating the calculation used to compute $S(\alpha, \lambda)$, one finds that for small λ, and α less than $\pi/4$, $T(\alpha, \lambda)$ is positive down to the limit $\lambda = 0$, with the expression for $T(\alpha, 0)$ being similar to that for $S(\alpha, 0)$. However, for the angle function $\theta(\alpha, \lambda)$ we have the identity $\theta(\alpha, \lambda) = 2(2\lambda)^{1/2}T(\alpha, \lambda)$. This gives us the following:

(a) $\theta(\alpha, \lambda)$ is positive for λ positive;

(b) $\theta(\alpha, \lambda)$ approaches 0 as λ approaches 0.

Since X is a connected set with more than one point (see Figure 3), it follows by continuity that on the set X the function $\theta(\alpha, \lambda)$ takes on a continuum of values $[0, \varepsilon]$ where ε is positive. Whenever $\theta(\alpha, \lambda)$ is a rational multiple of 2π the surface will close up. This establishes the existence of a countable number of isometrically distinct immersions of a torus into R^3 with constant mean curvature.

A View of the Immersed Tori

Let $\Omega = \Omega_{AB}$ be a representative rectangle chosen so that the smallest eigenvalue of the Laplace differential equation

$$\Delta v + \gamma v = 0 \text{ on } \Omega, v = 0 \text{ on boundary } \Omega \tag{9}$$

is $\gamma_1 = 1$. This means that $1 = \gamma_1 = \pi^2((1/A^2) + (1/B^2))$ and in particular A and B are both greater than π. We are to solve the differential equation

$$\Delta W + 2\lambda \sinh W \cosh W = 0 \text{ on } \Omega, W = 0 \text{ on } \delta\Omega. \tag{10}$$

We have the following facts regarding solutions to the differential equation (10).

(a) There exists a branch of positive solutions to (10) which bifurcate from the zero solution at $2\lambda = \gamma_1 = 1$ or $\lambda = \frac{1}{2}$.

(b) For any positive solution (W, λ) we must have $0 < \lambda < \frac{1}{2}$, and for any λ in this interval there exists at least one positive solution.

(c) As λ approaches 0 there is a curve of large positive solutions (W, λ) obtained by applying Theorem 3.

It is tempting to conjecture that the branch bifurcating from the zero solution at $\lambda = 1/2$ connects up with the branch of large solutions established in Theorem 3. In fact this is the case. It has recently been observed by U. Abresch [1] that there is a branch of positive solutions to the D.E. (10) on rectangular domains and with vanishing boundary values which can be solved in closed form using elliptic integrals. This branch connects the large solutions with small λ to the small solutions where $\lambda \to \lambda_1$. It is quite possible that all positive solutions lie on this branch. An alternate method is described in the book by G.L. Lamb [8] for example.

For each solution of the system (10) we may apply (7) to get a solution $\omega(u, v)$ to the differential equation (3) and then apply our recipe to construct an immersion $\bar{x}(u, v)$ with constant mean curvature. In the limit case where $W = 0$ the resulting immersion is simply a conformal mapping of the plane onto a circular cylinder whose cross section is a circle of radius one.

In the figures that follow we shall sketch the image $\bar{x}(u, v)$ of a portion of the fundamental rectangle $\sqrt{2\lambda}\Omega_{AB}$ as indicated in the first figure and labeled

$\{1, 2, 3, 4, 5, 6\}$. $A + $ sign indicates that $\omega(u, v)$ is positive and hence the Gauss curvature of the image surface $K = e^{-2\omega} \sinh \omega \cosh \omega$ is positive, while a $-$ sign indicates that both functions are negative. The rest of the surface is obtained by rotating the surface $180°$ about the normal line at the image of 2 followed by a series of reflections about the appropriate planes.

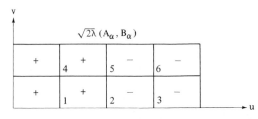

FIGURE 4. The fundamental domain $\sqrt{2\lambda}\Omega(\alpha) = \Omega(\alpha, \lambda)$.

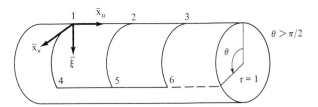

FIGURE 5. Case 1; $W = 0$, a pure cylinder.

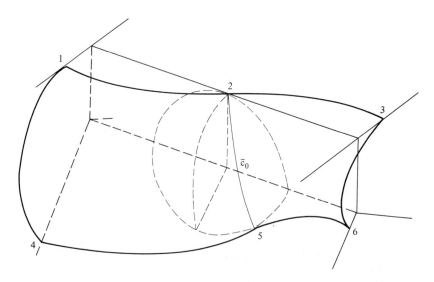

FIGURE 6. Case 2; W is positive on $\Omega(\alpha, \lambda)$ but not too large.

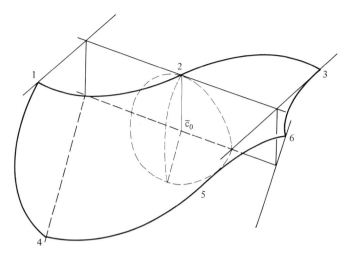

FIGURE 7. Case 3; W somewhat larger, the planes π_0, π_1 still separated.

If one keeps α fixed and lets λ approach 0, then one can easily show the following.

(1) $\int_{\bar{x}(\Omega^+)} K \, dA =$ area of the Gauss map $\to 4\pi$ as λ approaches 0.
(2) $\int_{\Omega^+} e^{2\omega} \, du \, dv =$ area of $\bar{x}(\Omega^+) \to 2\pi(2)^2$ as λ approaches 0.
(3) $\int_{\Omega^-} e^{2\omega} \, du \, dv =$ area of $\bar{x}(\Omega^-) \to 0$ as λ approaches 0.

These calculations suggest that as λ approaches 0, $\bar{x}(\Omega^+)$ takes on the shape of a sphere of radius 2.

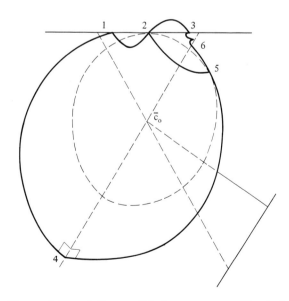

FIGURE 8. Case 4; the parallel planes π_0, π_1 are identical.

If one reflects the sketched Figure 8 about the plane of the paper ($\Pi_0 = \Pi_1$), you obtain a surface which resembles a clam shell. Upon rotating this shell $180°$ about the vertical line $c_0 - (2)$ one obtains the other shell. The combined figure is now a clam with the shells opened a bit.

Acknowledgments. We are fortunate to be able to include some color illustrations of the immersed torus. They were produced using computer techniques by David Hoffman and James T. Hoffman of the University of Massachusetts. Joel Spruck, Alexander Eydeland, and Michael Callahan were important contributors to this project. This work was supported in part by a grant from the National Science Foundation.

References

1. U. Abresch, S.F.B. Preprint, University of Bonn, Germany.
2. A.D. Alexandroff, *Uniqueness Theorems for Surfaces in the Large*, V. Vestnik, Leningrad Univ. No. 19 (1958) 5–8: Am. Math. Soc. Transl. (Series 2) **21**, 412–416.
3. L.P. Eisenhart, *A Treatise on the Differential Geometry of Curves and Surfaces*, Dover Reprint, 1960.
4. B. Gidas, W. Ni, and L. Nirenberg, *Symmetry and Related Properties via the Maximum Principle*, Comm. Math. Physics **68** (1979), No. 3, 209–243.
5. H. Hopf, *Differential Geometry in the Large*, (Seminar Lectures, New York Univ. 1946 and Stanford Univ. 1956) Lecture Notes in Mathematics No. 1000, Springer-Verlag, New York, 1983.
6. Wu-Yi Hsiang, *Generalized Rotational Hypersurfaces of Constant Mean Curvature in the Euclidean Space I*, Jour. Diff. Geometry **17** (1982), 337–356.
7. J.H. Jellett, *Sur la Surface dont la Courbure Moyenne est Constante*, J. Math. Pures Appl., **18** (1853), 163–167.
8. G.L. Lamb, *Elements of Soliton Theory*, Wiley-Interscience, 1980.
9. J.L. Moseley, *On Asymptotic Solutions for a Dirichlet Problem with an Exponential Singularity*, Rep Amr I, West Virginia University, 1981.
10. V.H. Weston, *On the Asymptotic Solution of a Partial Differential Equation with an Exponential Nonlinearity*, SIAM J. Math Anal **9** (1978), 1030–1053.
11. H.C. Wente, *Counterexample to a Conjecture of H. Hopf*, Pac. J. of Math. **121**, No. 1 (1986), 193–244.

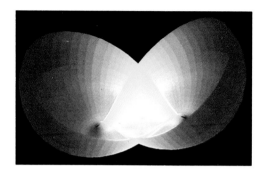

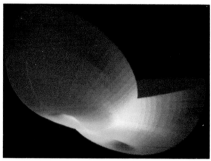

Basic annular building block of the surface.

Basic annular building block—second view.

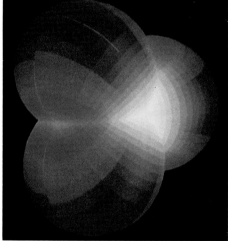

Basic building block joined to a reflected copy of itself.

The entire surface composed of three basic building blocks.

Three-lobed immersed torus of constant mean curvature.

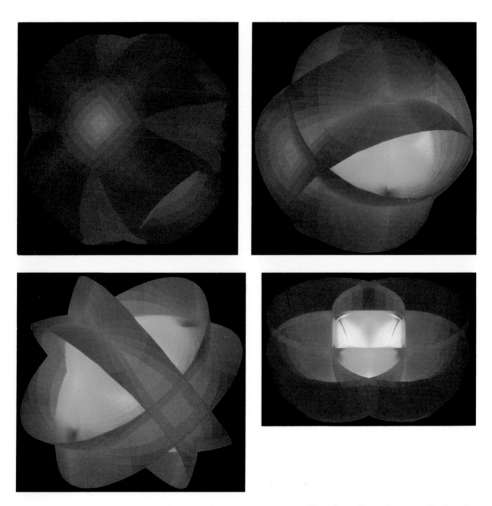

Four-lobed immersed torus of constant mean curvature. The first three images depict the surface with exterior portions removed. In each successive image, more is peeled away and the remaining piece enlarged. The fourth image depicts half the surface, produced by slicing through the reflective plane of symmetry orthogonal to the axis of rotation.

All of the images were generated by computer at the University of Massachusetts at Amherst using VPL, a graphics environment created by James T. Hoffman. The hardware was a Ridge 32 computer and a Raster Technologies Model One/380 graphics controller. ©1985, David Hoffman and James T. Hoffman.

The surface.

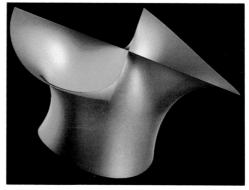

Portion of the surface in the half-space $x_2 \geq 0$. This corresponds to the region of M over the half-plane Im $(z) \geq 0$.

Portion of the surface in the half-space $x_3 \leq 0$. This corresponds to the region of M below the disk $|z| \leq 1$.

Complete embedded minimal surface of genus one with three ends.

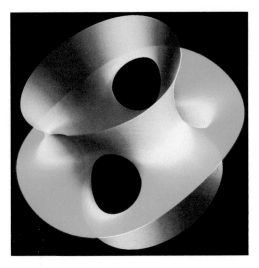

Genus two.

Genus three.

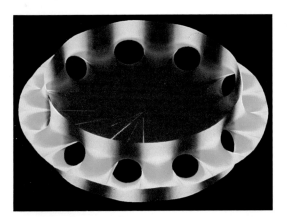

Genus nine.

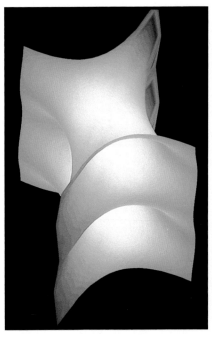

Portion of Scherk's second surface. A limiting ▷
surface for this family.

Higher genus complete embedded minimal surfaces with three ends.

All of the images were generated by computer at the University of Massachusetts at Amherst using VPL, a graphics environment created by James T. Hoffman. The hardware was a Ridge 32 computer and a Raster Technologies Model One/380 graphics controller. ©1985, David Hoffman and James T. Hoffman.

The Construction of Families of Embedded Minimal Surfaces

David. A. Hoffman

In this chapter I would like to briefly describe some new results in the classical theory of minimal surfaces. These discoveries represent joint work with William H. Meeks III. Our research made critical use of the graphics programming software developed by James T. Hoffman at the University of Massachusetts. The central theorem is the following existence result [4], [5], [6].

Theorem I. *For every $k \geqslant 1$, there exists a complete embedded minimal surface of finite total curvature $C(k) = -4\pi(2 + k)$ with three ends. It has the following properties.*

1. *Its symmetry group is the dihedral group $D(2k + 2)$ with $4(k + 1)$ elements generated by the orthogonal motions*

$$
K = \begin{bmatrix} 1 & 0 & 0 \\ 0 & -1 & 0 \\ 0 & 0 & 1 \end{bmatrix} \quad and \quad B = \begin{bmatrix} \cos\theta, & -\sin\theta, & 0 \\ \sin\theta, & \cos\theta, & 0 \\ 0, & 0, & -1 \end{bmatrix}, \quad \theta = \pi/(k + 1).
$$

   It is the unique embedded minimal surface of this genus with three ends, finite total curvature $C(k)$, and symmetry group $D(2k + 2)$.
2. *It consists of $4(k + 1)$ congruent pieces, each of which is a graph. The pieces are congruent under the action of $D(2k + 2)$.*
3. *It contains $k + 1$ coplanar straight lines crossing at equal angles at the unique fixed point of $D(2k + 2)$.*

The reader is referred to the computer-generated color plates, which accompany this paper.

History of the Problem

To explain the significance of the existence of these surfaces and their unusual properties, let me begin by giving a sequence of equivalent characterizations of a minimal surface. This will allow me to place the results in their proper historical context.

Definition I. Each point on the surface has a neighborhood which has least area among all surfaces with the same boundary.

The flat plane was the first example of a minimal surface. The second one was discovered by Euler. In the 1740s Euler posed and solved the following problem. Consider a surface of revolution defined by a profile curve $y = f(x)$, $a \leqslant x \leqslant b$, with the values fixed at a and b. Let $A(f)$ be the area of the surface. Which functions f are critical in the sense that the first derivative of A with respect to f is zero? The answer turns out to be that $f(x)$ must have a graph which is a catenary: $f(x) = \cosh(x)$. The surface of revolution is the well-known catenoid.

Definition II. If the surface is written locally as the graph of a real-valued function $z = f(x, y)$, then f satisfies the following (nonlinear elliptic) partial differential equation:

$$f_{xx}(1 + f_y^2) - 2f_x f_y f_{xy} + f_{yy}(1 + f_x^2) = 0. \tag{1}$$

This is in fact the Euler–Lagrange equation for the area functional of a graph, first established (not surprisingly) by Lagrange in the 1760s. The problem of minimizing area is one of the oldest problems in the Calculus of Variations.

In the 1770s, Meusnier gave the following geometric interpretation of equation (1).

Definition III. The mean curvature H, which is the average of the principal curvatures of the surface, vanishes identically:

$$H = 0. \tag{2}$$

Meusnier also showed that the helicoid, which is the image of R^2 under the mapping $X(t, s) = (t\cos(s), t\sin(s), s)$, is a minimal surface. Besides the flat plane, it is the only ruled minimal surface.

The minimal surface equation (2) was difficult to solve in the 18th century. The helicoid and catenoid were found by assuming certain symmetries in the surface which reduced the problem to the solution of an ordinary differential equation. No other explicit solution was discovered, to my knowledge, until 1835 when Scherk found several different ones. This research won a prize from the Paris Academy. The best known surface in this family is named after its discoverer and can be described as the zero set:

$$e^{x_3} - (\cos(x_2)/\cos(x_1)) = 0.$$

Another member of this family is Scherk's Second Surface which consists of the zero set:

$$\sin(x_3) - \sinh(x_1)\sinh(x_2) = 0.$$

This surface looks something like two orthogonal planes whose line of intersection has been replaced by tunnels burrowing through in alternating directions. It is a regular surface, free of self-intersections. It should be better known than it is. One of the reasons for this is its connection with Theorem I. If the examples in Theorem I are normalized to have Gauss curvature no bigger than one in absolute value, then they converge, as k goes to infinity, to Scherk's Second Surface.

While few new solutions to the minimal surface equation were found in the first half of the 19th century, the connection between minimal surfaces and harmonic functions was made.

Definition IV. The component functions, x_i, $i = 1, \ldots, 3$, of a minimal immersion X are harmonic functions. (They satisfy the Laplace–Beltrami equation.)

(Of course, this is an anachronism. What was known was that, in conformal coordinates, the component functions satisfied the ordinary Laplace equation.)

Some progress was made, using this insight, in integrating the minimal surface equation using analytic functions. Attention was focused upon the Plateau Problem (named after J.F. Plateau, the Belgian experimentalist who did fundamental research on problems involving surface tension and thin films): find the surface of minimum area spanning a given contour. Soap films, the kind that can be produced by dipping a wire frame into a solution of dishwashing liquid, are stable when their surface area cannot be reduced by small perturbations of the film. This is a consequence of the following fact: Both sides of the soap film are subject to the same pressure, and the pressure P, due to surface tension, which is normal to a surface interface is given by the formula $PdA = 2\sigma H dA$, where σ is the surface tension. We must have $H = 0$ because $P = 0$. Thus soap films spanning wire contours are actually minimal surfaces in the sense of all our previous definitions and also are really (at least relative) minima in a global sense. We could modify Definition I to read as follows.

Definition V (Experimentalist's definition). A sufficiently small piece of the surface is a mathematical realization of the soap film which forms on the contour coinciding with the boundary of the piece.

As late as the middle of the last century it was not possible to describe

analytically the solution to any boundary value problem for the minimal
surface equation unless the boundary curve lay on one of the few explicitly
known examples. Then in 1865, H.A. Schwarz announced the solution to the
problem of finding the surface of least area whose boundary was a regular
polygon consisting of four edges lying on the surface of a cube. (For more
details, see the beautiful book of Hildebrandt and Tromba [3].) This solution
was also found by Riemann several years earlier but published only after his
death. Schwarz, Riemann, Weierstrass, and others were able to solve the
boundary-value problem for a variety of polygonal boundaries. The critical
tool in this work was complex function theory. There is a deep connection
between minimal surfaces, analytic function theory, and the Gauss normal
mapping. (For a regular oriented surface in R^3, the map which associates to
a point on the surface its oriented unit-normal, considered as a point in S^2, is
called the Gauss normal map.)

Definition VI. The Gauss normal map G is anticonformal: wherever it is
nonsingular, it preserves the magnitude of angles while reversing their
orientation.

On a surface in R^3 we can always introduce, locally, conformal coordinates
(u_1, u_2). By introducing the local parameters $z = (u_1 + iu_2)$, the surface becomes a Riemann surface. If we compose the Gauss map G with stereographic
projection σ, the resulting map, $g = \sigma G$, from the Riemann surface to the
extended complex plane is, according to Definition VI, meromorphic precisely
on minimal surfaces. Conversely, given a meromorphic function, g, on a simply
connected domain in C, it is always the Gauss map of a minimal immersion
of the domain into R^3. This minimal surface is not unique. In fact, one can
specify an arbitrary analytic function $f \not\equiv 0$, on the domain and the pair will
determine a branched minimal surface up to translation.

The conformal minimal immersion is given by

$$X(z) = \text{Re} \int_{z_0}^{z} \Phi, \quad \text{where } \Phi = (f(1 - g^2), if(1 + g^2), 2fg)\, d\zeta. \quad \text{(EW1)}$$

This is the famous Enneper–Weierstrass representation formula, which allows
one to write down great numbers of minimal surfaces. It was an equivalent
formulation of this representation that was used by Schwarz to solve the
boundary-value problem. One can write the Schwarz surface as the minimal
surface with $f(z) = z$ and $g(z) = (1 - 14z^4 + z^8)^{-1}$ on an appropriate domain
in the complex plane.

The representation theorem was used by Riemann, Schwarz, and others
to build triply-periodic complete minimal surfaces out of basic pieces with
boundaries consisting of straight lines or plane curves. By arranging the
boundaries of the basic piece correctly, it could be insured that the resulting
minimal surface was free of self-intersections.

In recent years, new embedded complete periodic minimal surfaces have been discovered by A. Schoen and by Bill Meeks III, all constructed out of basic building blocks. However, this procedure can never produce a complete embedded minimal surface which is homeomorphic to a compact surface with a finite number of points removed. In fact, the only classical examples of complete embedded minimal surfaces with finitely-generated topology are the plane, the helicoid, and the catenoid. Topologically (and conformally), the plane and the helicoid are both once-punctured spheres and the catenoid is a twice-punctured sphere. To complete our historical background, it is necessary to weave in one more thread: the work of Osserman on complete minimal surfaces of finite total curvature.

The Gauss curvature K is the product of the principal curvatures, so by Definition II, $K \leqslant 0$ on a minimal surface. Since K is also the determinant of the differential of the Gauss map, the area of the Gaussian image is the total curvature of the surface. If the total curvature is finite, the differential of the Gauss map, which is meromorphic, must go to zero in some controlled way as one goes off to the boundary of the underlying Riemann surface. In fact, more is true.

Theorem (Osserman). *Let S be a complete minimal surface in R^3. Then S has finite total curvature if and only if*

(1) *S is conformally equivalent to a compact Riemann surface M of genus k with $r > 0$ points removed, and*
(2) *the Gauss map of S extends to a meromorphic function on M.*

In particular, if a complete minimal surface has finite total curvature, it has finite topology. It follows from Statement 2 that the total curvature $C(S)$ is an integer multiple of 4π. Osserman went on to show that

$$C(S) \leqslant -4\pi(k + r - 1), \qquad (3)$$

which is an improvement, in the minimal surface case, of the Cohn–Vossen estimate for the total curvature of a complete surface. He also observed that the Enneper–Weierstrass Representation makes sense on a Riemann surface. Here, instead of the auxiliary analytic function, f, one has to specify a holomorphic one-form η. What is needed in order to make sure that the mapping X in (EW1) is well-defined is that the integrals involved have no real periods. (For regularity, it is necessary and sufficient that the poles of g, of order m, coincide precisely with the zeros of η, of order $2m$.)

$$X(z) = \operatorname{Re} \int_{z_0}^{z} \Phi, \qquad \text{where } \Phi = ((1 - g^2), i(1 + g^2), 2g)\eta. \quad (\text{EW2})$$

In view of these results it is natural to replace the requirement of finite topology with the (stronger) assumption that the total curvature is finite. Such a surface could be constructed by specifying the Gauss map and auxiliary one-form, η, on a Riemann surface, then checking all the necessary conditions (including one, corresponding to completeness, that involves growth conditions on the integrands as one approaches the puncture points). However, it is easy to imagine that it would be quite difficult to check from these equations whether or not a given example was embedded.

The First New Example

The short list of embedded examples of complete minimal surfaces of finite topology that we have encountered so far—plane, catenoid, helicoid—constituted the complete list before we proved Theorem I. They have total curvatures equal to 0, -4π, and $-\infty$, respectively. In the category of finite total curvature we therefore had only the plane and the catenoid, both of genus $k = 0$, with $r = 1$ and $r = 2$ points removed, respectively. This is a rather thin class upon which to make conjectures. However, no new examples had been found since the 18th century, and some recent results indicated that there were obstructions to their existence. For example, Rick Schoen proved in 1982 that the catenoid was the unique embedded complete minimal surface with finite total curvature which was homeomorphic to a compact surface with two points removed [12]. There is no restriction on the genus. A little earlier, Meeks and Luquesio Jorge had proved that it was impossible to find a complete embedded minimal surface which had finite total curvature and was also homeomorphic to the sphere with 3, 4, or 5 points removed [8]. There had also been some false proofs that no examples other than the plane and the catenoid could exist.

Candidates could be screened by the following necessary conditions, first observed by Meeks and Jorge [8]. If the surface has genus k and r points removed then: (i) after a rotation of the sphere, the extended Gauss map G must be vertical at the ends; (ii) the total curvature must be equal to $-4\pi(k + r - 1)$. (This is the maximum value it can achieve according to (3)). In terms of the function g (which is the Gauss map in (EW2)), these requirements dictate that $g = 0$ or ∞ at the puncture points and the order of g is equal to $k + r - 1$. One such example was found by C. Costa in his thesis [1]. It had genus $k = 1$ and $r = 3$ ends. We investigated this surface using computer graphics and became aware of its marvelous symmetry. This symmetry was the key to establishing that the surface was embedded.

The process of understanding how to construct higher genus examples actually led to a simplification of the construction of the genus one example. We present a sketch of that here. (In our paper [5], the genus one example is explained in terms of the Weierstrass P-function on the square torus, as originally constructed by Costa.)

The Higher Genus Examples

Consider the punctured Riemann surface T, produced by removing the points $P_\infty = (\infty, \infty)$, $P_{-1} = (-1, 0)$, and $P_1 = (1, 0)$ from $\{(z, w) | w^2 = z(z^2 - 1)\}$. This is conformally a square torus minus three points. Moreover, the meromorphic functions z and w produced by projection onto the complex coordinate-planes are equal—up to a real multiplicative constant—to the Weierstrass P-function and its derivative, P'. We may construct a minimal immersion of T, using the Enneper–Weierstrass Representation, with $g = c/w$ and $\eta = (z/w)\,dz$. (Here c is a real constant determined by the requirement that there be no real periods in the Enneper–Weierstrass Representation.) To get a simple picture of this surface, it is useful to make an elementary transformation in the z plane. Let $\zeta = (z + 1)/(z - 1)$. The resulting identity which defines the surface is (replacing ζ by z) $w^2 = 4z(z + 1)/(z - 1)^3$, and the omitted points are now $Q_0 = (0,0)$, $Q_\infty = (\infty, \infty)$, and $Q_1 = (1, \infty)$. The Gauss map will be $g = c/w$ and $\eta = (w/z)\,dz$, presuming that we integrate from $z_0 = Q_{-1} = (-1, 0)$. The points Q_0 and Q_∞ correspond to the ends with catenoid-like behavior, while Q_1 is the flat end in the middle. Substitution in the Enneper–Weierstrass formula will show that the third integrand is $2c\,dz/z$. Hence, the third component of the immersion X is $x_3 = 2c \ln |z|$. The points Q_0 and Q_∞ correspond to the ends with catenoid-like growth, while Q_1 corresponds to the flat end in the middle.

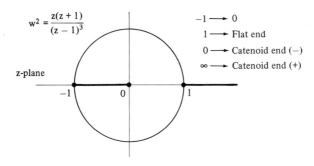

FIGURE 1.

Consider the surface to be a double-sheeted covering of the extended complex z-plane with the branch cuts as indicated in Figure 1. The curves lifting the unit circle have $x_3 = 2c \ln |1| = 0$, so they are mapped into the plane $x_3 = 0$. They turn out to be straight lines emanating from the origin in R^3, which is the image of the point $Q_{-1} = (-1, 0)$. Similarly, each circle $|z| = \varepsilon$ in the z-plane lifts to a curve lying in the plane $x_3 = 2c \ln(\varepsilon) = $ constant. This means that the region on the Riemann surface above the unit disk is mapped onto the region of the minimal surface below the plane $x_3 = 0$; the covering of the exterior of the unit disk is mapped onto the region of the surface above the $x_3 = 0$ plane.

The inversion $z \to 1/z$ lifts to a conformal diffeomorphism of the surface that is actually an isometry (induced by the symmetry KB of the surface given by Theorem I). Complex conjugation $z \to \bar{z}$ lifts to an isometry of the surface that is actually induced by reflection K. See the computer-generated pictures at the beginning of the chapter.

To produce the higher genus examples, start with the Riemann surface

$w^{k+1} = z^k(z^2 - 1)$. Let the Gauss map be $g = c/w$ and the auxiliary one-form be $\eta = (z/w)^k \, dz$. (c is a real constant determined by the need to kill off the periods.) The same linear fractional transformation of the z-coordinate that we used above will allow one to view the surface as a $(k + 1)$-sheeted covering of the z-plane with the same branch points and cuts as in the genus-one example. Moreover, the third component of the immersion is again equal to $2c \log |z|$. The unit circle in the z-plane lifts to $k + 1$ straight lines meeting at Q_{-1} and diverging to the flat "planar" end at Q_1. The mappings $z \to 1/z$ and $z \to \bar{z}$ lift to isometries of the surface which are the restriction of symmetries of the surface: $z \to 1/z$ corresponds to KB in Theorem I; the map $z \to \bar{z}$ corresponds to K in Theorem I.

Not only does this way of constructing the example simplify the proof of embeddedness, but it also provides a simple means of drawing the surface. First of all, we need only draw the half of the surface that lies over the unit disk; the other piece can be produced by rotation and reflection. Moreover, we can draw the curve on the surface over $|z| = \varepsilon$ by first lifting it to the surface and then integrating the first two integrands in the Enneper–Weierstrass formula along this lift. It doesn't matter where we start; since the resulting curve must be symmetric with respect to the origin, we can simply make its average value equal to zero. The one problem with this procedure is that as ε approaches 1, the lifted curve approaches a collection of curves on the surface that get mapped into divergent lines. Thus the length of the lift blows up. Some adjustments must be made to avoid this computational obstacle.

More Results

As mentioned above, Schoen [12] has proved that the only embedded minimal surface of any genus with finite total curvature and two ends ($r = 2$) is the catenoid. The catenoid has genus $k = 0$, two ends, and total curvature -4π. Meeks and Jorge [8] proved that there is no complete embedded minimal surface of finite total curvature with genus $k = 0$ and 3, 4, or 5 ends. Since embeddedness implies that the total curvature must equal $-4\pi(k + r - 1)$, where r is the number of ends, it follows from these results and Theorem I that:

Corollary I. *There exists a complete embedded minimal surface of total curvature* $-4\pi j$ *for every positive integer* j, *except* $j = 2$.

We have also established the existence of one-parameter families of minimal surfaces.

Theorem II. *Each of the examples in Theorem I sits in a 1-parameter family of embedded minimal surfaces.*

The other members of the 1-parameter family have half the symmetry of the examples in Theorem I. The next result is a first step toward generalizing our work to complete embedded minimal surfaces with infinite total curvature.

Theorem III. *A properly embedded minimal surface can have, at most, two annular ends with infinite total curvature.*

Theorem IV. *A properly immersed minimal surface cannot lie in a half space. More generally, two properly immersed minimal surfaces which never intersect must be parallel planes.*

To each end of a complete minimal surface with finite total curvature one can associate an integer which measures the winding of the surface at that end. The sum of these integers we shall denote by $n(M)$. The total curvature can then be written as $-4\pi(k - 1 + r) - 2\pi(n(M) - r)$, where k is the genus. Also, $n(M) = r$, where r is the number of ends, if and only if each end is separately embedded.

Theorem V. *A complete connected minimal surface M of finite total curvature cannot intersect itself with multiplicity more than $n(M) - 1$. In case M has r ends which are all embedded, it cannot intersect itself with multiplicity greater than $r - 1$.*

Acknowledgment. This research was partially supported by NSF Grant MCS-8301936.

References

[1] C. Costa, *Imersões minimas completas em R^3 de gênero um e curvatura total finita*, Doctoral thesis, IMPA, Rio de Janeiro, Brasil, 1982. (Example of a complete minimal immersion in R^3 of genus one and three ends, Bull. Soc. Bras. Mat., **15** (1984) 47–54.)

[2] L. Barbosa and G. Colares, *Examples of Minimal Surfaces* (to appear in Springer Lecture Notes series).

[3] S. Hildebrandt and A.J. Tromba, *The Mathematics of Optimal Form*, Scientific American Library, W.H. Freeman and Company, New York, 1985.

[4] D. Hoffman and W. Meeks III, *Complete embedded minimal surfaces of finite total curvature*, Bull. A.M.S., January 1985.

[5] D. Hoffman and W. Meeks III, *A complete embedded minimal surface with genus one, three ends and finite total curvature*, Journal of Differential Geometry, March 1985.

[6] D. Hoffman and W. Meeks III, *The global theory of embedded minimal surfaces* (in preparation).

[7] D. Hoffman, *The discovery of new embedded minimal surfaces: elliptic functions; symmetry; computer graphics*, Proceedings of the Berlin Conference on Global Differential Geometry, Berlin, June 1984.

[8] L. Jorge and W. Meeks III, *The topology of complete minimal surfaces of finite total Gaussian curvature*, Topology, **22** No. 2 (1983), 203–221.

[9] J.C.C. Nitsche, *Minimal surfaces and partial differential equations*, in Studies in Partial Differential Equations (Walter Littman, ed.), MAA Studies in Mathematics, Vol. 23, The Mathematical Association of America, 1982.

[10] R. Osserman, *A Survey of Minimal Surfaces*, 2nd Edition, Dover Publications, 1986.

[11] I. Peterson, *Three Bites in a Doughnut*, Science News, 3/16/85, 69–96.

[12] R. Schoen, *Uniqueness, symmetry, and embeddedness of minimal surfaces*, J. Diff. Geom. **18** (1983), 791–809.

Boundary Behavior of Nonparametric Minimal Surfaces—Some Theorems and Conjectures

Suppose D is a domain in the plane which is locally convex at every point of its boundary except possibly one, say $(0,0)$, and ϕ is continuous on ∂D except possibly at $(0,0)$, where it might have a jump discontinuity. Then for all directions from $(0,0)$ into D, the radial limits of f exist, where f is the solution of the minimal surface equation in D or of an equation of prescribed (bounded) mean curvature in D with $f \in C^0(\bar{D}\backslash\{(0,0)\})$ and $f = \phi$ on $\partial D\backslash\{(0,0)\}$. Some conjectures which would generalize this result are mentioned.

Suppose that D is a bounded domain in $\mathbb{R}^n$ and $f \in C^2(D)$ is a bounded solution of the minimal surface equation. How does f behave on ∂D? That f may not extend continuously onto $\bar{D}$ is shown by several examples ([9], [4], [8], [10]) and implied by certain theorems ([1], [2], and especially [5]). How, then, does f behave near a point of the boundary of D at which it is not continuous? If $n = 2$, the answer is known in some cases, as indicated below. Further, if the first conjecture below is correct, then the answer is known in all cases. On the other hand, if $n > 2$, nothing is known at this time.

Let D be a bounded, connected, simply connected, open subset of $\mathbb{R}^2$ such that ∂D is Lipschitz, $N = (0,0) \in \partial D$, and D is locally convex at each point of its boundary except possibly N. Suppose that $D = \{(r,\theta)|\alpha < \theta < \beta, 0 < r < r(\theta)\}$ with $-\pi < \alpha < 0 < \beta < \pi$, where (r,θ) represents polar coordinates about N. This last conditions implies that the x-axis divides ∂D near N into two components.

Definition. By $C^*(\partial D)$ we mean those functions $\phi \in C^0(\partial D\backslash\{N\})$ such that

$$\phi(N+) = \lim \phi(P) \text{ as } P \in \partial D \cap \{(x,y)|y > 0\} \text{ approaches } N,$$

and

$$\phi(N-) = \lim \phi(P) \text{ as } P \in \partial D \cap \{(x,y) | y < 0\} \text{ approaches } N$$

each exist.

Definition. For $\phi \in C^*(\partial D)$, define $f = f(\cdot, \phi)$ as the function in $BV(D)$ which minimizes

$$J(v) = J(v, \phi) = \iint_D \sqrt{1 + |Dv|^2} + \int_{\partial D} |v - \phi|$$

for $v \in BV(D)$.

We know that $f \in C^2(D) \cap C^0(\bar{D} \backslash \{N\})$ and $f = \phi$ on $\partial D \backslash \{N\}$. We will assume throughout that $f \notin C^0(\bar{D})$. We will let

$$S_0 = S_0(\phi) = \{(x, y, f(x, y)) | (x, y) \in D\},$$
$$\Gamma_0 = \Gamma_0(\phi) = \{(x, y, \phi(x, y)) | N \neq (x, y) \in \partial D\},$$

S equal the closure of S_0, Γ equal the closure of Γ_0, Γ^+ equal the closure of $\Gamma \cap \{(x,y) | y > 0\}$, and Γ^- equal the closure of $\Gamma \cap \{(x,y) | y < 0\}$. For $\alpha < \theta < \beta$, define

$$Rf(\theta) = \lim_{t \to 0^+} f(t\cos(\theta), t\sin(\theta))$$

if this exists. Set $Rf(\alpha) = \phi(N-)$ and $Rf(\beta) = \phi(N+)$.

In [7], which generalizes the result of [6], the following theorem is proven. This theorem is also true when f is a solution of an equation of prescribed (bounded) mean curvature with $f \in C^0(\bar{D} \backslash \{N\})$ and $f = \phi$ on $\partial D \backslash \{N\}$ (see [7]; also [3]).

Theorem. *If $\phi \in C^*(\partial D)$, then $Rf(\theta)$ exists for each $\theta \in [\alpha, \beta]$ and $Rf \in C^0([\alpha, \beta])$. Further, either*

Rf is weakly monotonic on $[\alpha, \beta]$, constant on $[\alpha, \alpha']$ and on $[\beta', \beta]$, and strictly monotonic on $[\alpha', \beta']$, for some $\alpha \leq \alpha' < \beta' \leq \beta$, or

Rf is constant on $[\alpha, \alpha']$, strictly increasing (decreasing) on $[\alpha', \theta_L]$, constant on $[\theta_L, \theta_L + \pi]$, strictly decreasing (increasing) on $[\theta_L + \pi, \beta']$, and constant on $[\beta', \beta]$, for some $\alpha \leq \alpha' < \theta_L$ and $\theta_L + \pi < \beta' \leq \beta$.

If Γ^- is a straight line segment in a neighborhood of $(N, \phi(N-))$ which meets the z-axis nontangentially, then $\alpha' = \alpha$ or $\alpha' = \alpha + \pi$; similarly for Γ^+ and β'.

The angles at which Γ^- meets the z-axis do not determine whether $\alpha' = \alpha$ or $\alpha = \alpha' + \pi$ as the following example shows.

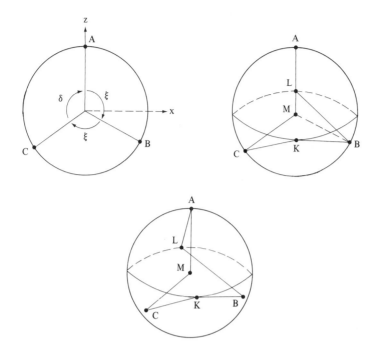

EXAMPLE (refer to the figures illustrated above). Let $\delta \in (0, \pi)$ and set $\xi = \pi - \delta/2$. Here we will have $\alpha = -\pi$, $\alpha' = 0$, $\beta = \pi/2$, and δ as the angle between Γ^- and the positive z-axis. Set $A = (0, 0, 1)$, $B = (\sin(\xi), 0, \cos(\xi))$, $C = (\sin(2\xi), 0, \cos(2\xi))$, $K = (0, 0, 1)$, $L = (0, -1, 0)$, and $M = (0, 0, 0)$. Let $Q1$ be the quadrilateral with successive vertices B, K, C, M and let $S1$ be the minimal surface spanning $Q1$. Since $Q1$ has a simple convex projection on the x-y plane, $S1$ is unique and is the graph of a function over the x-y plane. Now let $S2$ be the reflection of $S1$ across the line BM and set $S = S1 \cup S2$. The boundary of S is the polygon Γ with successive vertices M, A, L, B, K, C, M. The following lemma implies $S2 \backslash \Gamma$, and so $S \backslash \Gamma$ is the graph of a function over the x-y plane.

Lemma. *Suppose Γ is a Jordan curve in $\mathbb{R}^3$ which has a convex projection on the x-y plane which is a simple projection except at one point. Suppose that only one minimal surface (of the type of the disc) S spans Γ. Then $S \backslash \Gamma$ is the graph of a function over the x-y plane.*

PROOF. Define D as the open set bounded by the projection of Γ on the x-y plane. We may assume $N \in \partial D$ and Γ has a simple projection onto each point of $\partial D \backslash \{N\}$. Thus Γ contains a line segment over N. Define $\phi \in C^0(\partial D \backslash \{N\})$ so that the graph of ϕ is a subset of Γ. Then $\phi \in C^*(\partial D)$ and from Proposition 1 and Theorem 1 of [7], $S(\phi)$ spans Γ. Since only one minimal surface spans Γ, $S = S(\phi)$ and $S \backslash \Gamma$ is the graph of $f(\cdot, \phi)$ over D. $\qquad \square$

These results lead one to consider generalizations. The following two conjectures are generalizations which I believe are true but cannot yet prove. The first conjecture, if true, would mean that the example of Leon Simon ([10], Thm. 2) is well behaved in a certain sense.

Conjecture. *Let D be any bounded, connected, simply connected, open subset of $\mathbb{R}^2$ and let $f \in C^2(D)$ be a bounded solution of the minimal surface equation in D. Let p be any point of ∂D and let (r, θ) be polar coordinates about p. Then if $\{(r, \theta) | 0 < r < \varepsilon\} \subseteq D$ for some $\varepsilon > 0$,*

$$Rf(\theta) = Rf(\theta, p) = \lim_{r \to 0^+} f(r, \theta)$$

exists.

Conjecture. *Under the same conditions as above, except that f is a bounded solution of an equation of bounded mean curvature, radial limits as indicated above exist.*

In each of the cases above, our theorem or its extension to bounded solutions of equations of prescribed (bounded) mean curvature implies that if $\{(r, \theta) | \alpha < \theta < \beta, 0 < r < r(\theta)\} \subseteq D$ and $Rf(\alpha)$ and $Rf(\beta)$ each exist, then $Rf(\theta)$ exists for all $\theta \in (\alpha, \beta)$ and $Rf(\cdot, p) \in C^0([\alpha, \beta])$. A suitable nonparametric proof of the first conjecture might lead to the proof of the second and might suggest a proof of a positive answer to the following question.

Suppose D is a bounded, connected, open subset of $\mathbb{R}^n$ and f is a bounded solution of the minimal surface equation (or the equation of prescribed (bounded) mean curvature) in D. Suppose $p \in \partial D$ and $\omega \in S^{n-1}$ such that

$$\{p + r\omega | 0 < r < \varepsilon\} \subseteq D \quad \text{for some } \varepsilon > 0.$$

Then does

$$Rf(\omega, p) = \lim_{r \to 0^+} f(p + r\omega)$$

necessarily exist?

References

1. M. Beeson, *The Behavior of a Minimal Surface in a Corner*, Arch. Rat. Mech. Anal. **65**(4), (1977), 379–393.
2. G. Dziuk, *Über quasilinear elliptische Systeme mit isotherman Parametern on Ecken der Randkurve*, Analysis **1**, (1981), 63–81.
3. A. Elcrat and K. Lancaster, *Boundary Behavior of a Non-Parametric Surface of Prescribed Mean Curvature Near a Reentrant Corner* Tran. Amer. Math. Soc. **297** (1986), 645–650.
4. R. Finn, *Remarks Relevant to Minimal Surfaces, and to Surfaces of Prescribed Mean Curvature*, J. d'Anal. Math. **14**, (1965), 139–160.
5. H. Jenkins and J. Serrin, *The Dirichlet Problem for the Minimal Surface Equation in Higher Dimensions*, J. Reine Angew. Math. **229**, (1968), 170–187.

6. K. Lancaster, *Boundary Behavior of a Non-Parametric Minimal Surface in* $\mathbb{R}^3$ *at a Non-Convex Point*, Analysis **5**, (1985), 61–69.
7. K. Lancaster, *Nonparametric Minimal Surfaces in* $\mathbb{R}^3$ *whose Boundaries have a Jump Discontinuity* (to appear).
8. J.C.C. Nitsche, *On the Nonsolvability of Dirichlet's Problem for the Minimal Surface Equation*, J. Math. Mech. **14**, (1965), 779–788.
9. T. Radó, *Contributions to the Theory of Minimal Surfaces*, Acta. Litt. Scient. Univ. Szeged **6**, (1932), 1–20.
10. L. Simon, *Boundary Behavior of Solutions of the Non-Parametric Least Area Problem*, Bull. Austral. Math. Soc. **26**, (1982), 17–27.

On Two Isoperimetric Problems with Free Boundary Conditions

S. Hildebrandt

During the last years, free boundary problems for minimal surfaces have found much attention. Beautiful new existence results were discovered by Struwe [25], Grüter–Jost [11], and Jost [19]. The question of boundary regularity was discussed by Hildebrandt–Nitsche [14], [15], [16], [17], Grüter–Hildebrandt–Nitsche [9], Dziuk [3], [4], Küster [20], and Ye [26]. Various regularity theorems are optimal, although several questions are still open. A survey of some of these results can be found in [12].

In this chapter I shall report on two isoperimetric problems which are closely related to the previously mentioned work on minimal surfaces. The first is the so-called *thread problem*. This is concerned with surfaces of minimal area within a boundary configuration $\langle \Gamma, \Sigma \rangle$ that consists of a fixed rectifiable Jordan arc Γ and a movable arc Σ of fixed length, the *thread*, connecting the two endpoints P_1 and P_2 of Γ. An existence theorem was obtained by H.W. Alt [1], whereas J.C.C. Nitsche [22], [23] proved some regularity. He showed that the nonattaching part of Σ between singular points consists of regular curves of class C^∞. I shall report on an improvement of this result found by Dierkes–Hildebrandt–Lewy [2].

The second part of this chapter will be concerned with the question of boundary regularity of solutions for *partition problems*. Solutions of minimal area were already treated in the paper [18] by Hildebrandt–Wente. The general case of stationary solutions, which is to be reported, is dealt with in a forthcoming paper [10] by Grüter–Hildebrandt–Nitsche.

1. The Thread Problem

Before we consider the minimum problem described in the introduction, we shall investigate *stationary solutions* to the thread problem.

Let $B = \{w : |w| < 1\}$ be the open unit disk in the complex w-plane, and denote by r, Θ polar coordinates about the origin: $w = re^{i\Theta}$. A mapping $X: B \to \mathbb{R}^3$ is called a minimal surface if the equations

$$\Delta X = 0, \tag{1}$$

$$X_w \cdot X_w = 0, \tag{2}$$

hold on B. Equation (2) can be written in the form

$$r^2 |X_r|^2 = |X_\Theta|^2, \qquad X_r \cdot X_\Theta = 0. \tag{2'}$$

The Dirichlet integral $D_B(X)$ of a mapping $X \in H^{1,2}(B, \mathbb{R}^3)$ is given by

$$D_B(X) = \frac{1}{2} \int_B |\nabla X|^2 r \, dr \, d\Theta,$$

where $|\nabla X| = |X_r|^2 + r^{-2}|X_\Theta|^2$.

The following result due to F. and M. Riesz and to Tsuji is basic for our analysis (see [23], pp. 285–290).

Lemma 1. *Let* $X: B \to \mathbb{R}^3$ *a disk-type minimal surface, and denote by* X^*: $B \to \mathbb{R}^3$ *the adjoint minimal surface to* X *which, up to an additive constant, is uniquely determined by the equations*

$$X_r = \frac{1}{r} X_\Theta^*, \qquad \frac{1}{r} X_\Theta = -X_r^*.$$

Assume that $D_B(X) < \infty$ *and* $\int_{\partial B} |dX| < \infty$. *Then we have:*

(i) X *and* X^* *are of class* $C^0(\bar{B}, \mathbb{R}^3)$, *and*

$$D_B(X) = D_B(X^*), \qquad \int_{\partial B} |dX| = \int_{\partial B} |dX^*|.$$

(ii) *The boundary values* $X(1, \Theta)$ *and* $X^*(1, \Theta)$ *are absolutely continuous functions of* Θ, *and* $X_\Theta(r, \dot{\Theta})$, $X_\Theta^*(r, \Theta)$ *tend in the* L^2-*sense to the derivatives* $X_\Theta(1, \Theta)$, $X_\Theta^*(1, \Theta)$ *of the boundary values* $X(1, \Theta)$ *and* $X^*(1, \Theta)$, *respectively, as* $r \to 1 - 0$. *Moreover,* $X_r(r, \Theta)$ *and* $X_r^*(r, \Theta)$ *converge in* L^2 *to boundary values as* $r \to 1 - 0$, *and we set*

$$X_r(1, \Theta) = \lim_{r \to 1-0} X_r(r, \Theta), \qquad X_r^*(1, \Theta) = \lim_{r \to 1-0} X_r^*(r, \Theta).$$

It follows that a.e.

$$X_r(1, \Theta) = X_\Theta^*(1, \Theta), \qquad X_\Theta(1, r) = -X_r^*(1, \Theta),$$

$$|X_r(1, \Theta)| = |X_\Theta(1, \Theta)|, \qquad X_r(1, \Theta) \cdot X_\Theta(1, \Theta) = 0.$$

(iii) *If* C *is an open subarc of* ∂B, *and* ξ *is a test function of class* $H^{1,2}(B, \mathbb{R}^3) \cap L^\infty(C, \mathbb{R}^3)$ *with* $\xi = 0$ *on* $\partial B - C$, *then*

$$\int_B \nabla X \cdot \nabla \xi \, r \, dr \, d\Theta = \int_C X_r \cdot \xi \, d\Theta. \tag{3}$$

(iv) *If $X \not\equiv$ const on B, then $X_\Theta(1, \Theta)$ and $X_\Theta^*(1, \Theta)$ vanish at most on a subset of $[0, 2\pi]$ of one-dimensional measure zero.*

Definition. A minimal surface $X: B \to \mathbb{R}^3$ is said to be a *stationary solution of the thread problem with respect to some open subarc C of ∂B* if the following holds: (i) $D_B(X) < \infty$, $\int_{\partial B} |dX| < \infty$, $X(w) \not\equiv$ const; (ii) there is a real number $\lambda \neq 0$ such that

$$\int_B \nabla X \cdot \nabla \xi \, r dr \, d\Theta + \lambda \int_C \frac{X_\Theta}{|X_\Theta|} \cdot \xi_\Theta \, d\Theta = 0 \tag{4}$$

holds for all $\xi \in C^1(\bar{B}, \mathbb{R}^3)$ with $\xi = 0$ on $\partial B - C$.

Taking the identity (3) into account, we arrive at

$$\int_C (X_r \cdot \xi + \lambda |X_\Theta|^{-1} X_\Theta \cdot \xi_\Theta) \, d\Theta = 0.$$

On the other hand,

$$\int_C X_r \cdot \xi \, d\Theta = \int_C X_\Theta^* \cdot \xi \, d\Theta = -\int_C X^* \cdot \xi_\Theta \, d\Theta.$$

Thus (4) is equivalent to

$$\int_C (X^* - \lambda |X_\Theta|^{-1} X_\Theta) \cdot \xi_\Theta \, d\Theta = 0 \tag{5}$$

for all $\xi \in C^1(\bar{B}, \mathbb{R}^3)$ with $\xi = 0$ on $\partial B - C$.

DuBois–Reymond's lemma now implies that (5) and therefore also (4) are equivalent to the following property of X:

There exists a constant vector $P \in \mathbb{R}^3$ such that

$$X^* = \lambda |X_\Theta|^{-1} X_\Theta + P \qquad \text{a.e. on } C \tag{6}$$

holds. Now we can prove the following.

Theorem 1. *Let $X: B \to \mathbb{R}^3$ be a minimal surface which is a stationary solution of the thread problem with respect to the open arc $C \subset \partial B$. Then, for some $P \in \mathbb{R}^3$ and some $\lambda \in \mathbb{R}$, $\lambda \neq 0$, equation (6) is satisfied. Moreover, X and its adjoint X^* are real-analytic on $B \cup C$, and X^* intersects the sphere*

$$S = \{Z \in \mathbb{R}^3 : |Z - P|^2 = \lambda^2\}$$

orthogonally along its free trace $\Sigma^* = \{X^*(w): w \in C\}$. Both X and X^* have no boundary branch points of odd order on C. Finally, $\Sigma = \{X(w): w \in C\}$ has a representation $\chi(s)$, $0 < s < 1$, by its arc length s as parameter, which is of class C^2 and satisfies $|\dot{\chi}(s)| \equiv 1$, $|\ddot{\chi}(s)| \equiv 1/|\lambda|$. Thus Σ represents a regular curve of constant curvature $\kappa = 1/|\lambda|$.

PROOF. As we have noticed, the assumption on X implies that (6) holds for some $P \in \mathbb{R}^3$ and some $\lambda \in \mathbb{R}$, $\lambda \neq 0$. Taking the continuity of $X^*(1, \theta)$ into account, we infer that

$$|X^* - P|^2 = \lambda^2 \quad \text{on } C. \tag{7}$$

In other words, the trace Σ^* lies on S. Moreover, we also obtain

$$X^* - P = -\lambda |X_r^*|^{-1} \cdot X_r^* \quad \text{a.e. on } C. \tag{8}$$

That is, X_r^* is normal to S a.e. on C. Thus, X^* has for almost all $w \in C$ a tangent plane that is meeting S at a right angle. By straightforward reasoning, we conclude that the adjoint surface X^* is a stationary minimal surface within the boundary configuration $\langle \Gamma^*, S \rangle$ consisting of the arc $\Gamma^* = \{X^*(w): w \in \partial B - C\}$ and of the surface S; see [9] and [20]. We therefore can apply Theorem 1 of [9] to X^* and obtain that X^* can be analytically continued across C as a minimal surface. We then infer that both X and X^* are real-analytic in $B \cup C$, as we have claimed. Actually this result can also be derived by the methods of [21] since we know that X and X^* are absolutely continuous along ∂B.

The other statements of the theorem follow from a detailed analysis of the mappings X and X^* at boundary branch points $w_0 \in I$. For details, we refer the reader to the paper [2] by Dierkes–Hildebrandt–Lewy, from which the previous discussion has been taken.

We note that, by appropriate examples, we can demonstrate that *stationary solutions of the thread problem* in general will have branch points of even order. Thus Theorem 1 cannot be improved. However, for real minima of the area, a stronger result holds.   □

Theorem 2. *Let* $X: B \to \mathbb{R}^3$ *be a minimal surface that minimizes Dirichlet's integral within the class of* $H^{1,2}$-*surfaces which map* $C^- = \partial B \cap \{\text{Im } w \leqslant 0\}$ *continuously and in a weakly monotonic way onto the arc* Γ, *and have a total variation* $\sigma = \int_{C^+} |dX|$ *on* $C^+ = \partial B - C^-$ *with* $|P_1 - P_2| < \sigma < \text{length of } \Gamma$. *Then* $X(w)$ *can analytically be continued as a minimal surface across the arc* C^+, *and it has on* C^+ *neither branch points of odd order nor true branch points of even order. If, moreover, the boundary mapping* $X: \partial B \to \mathbb{R}^3$ *is assumed to be an embedding, then* $X(w)$ *has also no false branch points of even order on* C^+. *Correspondingly, the free trace* $\Sigma = \{X(w): w \in C^+\}$ *is a regular, real-analytic curve of constant curvature.*

Remarks to the Proof of Theorem 2. One starts by establishing the existence of a Lagrange parameter λ. This is not totally trivial since the applicability of the standard Lagrange multiplier theorem is not guaranteed. Details can be found in Lemmas 2–4 of [2].

The absence of boundary branch points on C^+ follows essentially from a well-known theorem by Gulliver and Lesley, as is pointed out at the end of [2].

2. The Partition Problem

Let $\mathscr{K}$ be a bounded convex body in $\mathbb{R}^3$ with boundary T. We consider the following *partitioning problem* $\mathscr{P}$:

Determine a rectifiable surface $\bar{S}$ of minimal or stationary area, with boundary Σ contained in T, which divides $\mathscr{K}$ into two parts $\mathscr{K}_1$ and $\mathscr{K}_2$ such that

$$\text{meas } \mathscr{K}_1 = \sigma \text{ meas } \mathscr{K}, \qquad \text{meas } \mathscr{K}_2 = (1 - \sigma)\text{meas } \mathscr{K},$$

where σ denotes a preassigned constant with $0 < \sigma < 1$.

Applying the work of De Giorgi, one can obtain sets of finite perimeter as minimal solutions of the afore stated partition problem, and interior regularity follows from results by Gonzalez–Massari–Tamanini [6], and Giusti [5].

Boundary regularity for minimal partitioning surfaces was first verified by Hildebrandt–Wente [18]. The stationary case was treated by Grüter–Hildebrandt–Nitsche in [10]. We shall report on the results of this paper.

One first realizes that every solution $\bar{S}$ of $\mathscr{P}$ is a surface of constant mean curvature (in special situations: a minimal surface), perpendicular to the boundary T. Interior regularity follows from Grüter's work [7, 8]. Thus the emphasis of [10] lies in the proof of boundary regularity for $\bar{S}$, and it will already be assumed that the interior part S of $\bar{S}$ is a regular C^1-manifold which divides int $\mathscr{K} - S$ into two disjoint parts of preassigned measure. As a consequence, S has a conformal parameter representation on some Riemann surface B. Generally speaking, the surface S need not be of disk-type. It could, for instance, also have the topological type of an annulus, as is the case for a cylinder, a catenoid, or an arbitrary Delaunay surface. Although the approach of [10] can easily be modified to handle this case or more general cases as well, we shall restrict ourselves here to solutions of $\mathscr{P}$ which can be parametrized over a disk.

Let us fix the assumptions needed for the proof of the subsequent boundary regularity theorem.

Assumption (A1).

(i) S has a conformal parameter representation $X: B \to \mathbb{R}^3$ of class $C^1(B, \mathbb{R}^3)$ on the unit disk B (that is, (2) holds).

(ii) S has finite area, so that its representation X is of class $H^{1,2}(B, \mathbb{R}^3)$. We assume that $\Sigma := X(\partial B)$ is contained in T, where the boundary condition $\Sigma \subset T$ is to be interpreted in the sense that the L^2-trace of X on ∂B lies in T.

(iii) The surface $S = X(B)$ omits a neighborhood of some point $p \in T$.

(iv) The map $X: B \to \mathbb{R}^3$ is an embedding of B into the interior of $\mathscr{K}$ such that

$$\text{int } \mathscr{K} - S = \Omega_1 \cup \Omega_2, \qquad \Omega_1 \cap \Omega_2 = \varnothing,$$

where Ω_1 and Ω_2 are simply-connected regions and

$$\text{meas } \Omega_1 = \sigma \text{ meas } \mathscr{K}, \qquad \text{meas } \Omega_2 = (1 - \sigma)\text{meas } \mathscr{K}.$$

The main result of [10] is then the following.

Theorem 3. *Let T be a regular surface of class* C^3, *or* $C^{m,\alpha}(m \geqslant 3, 0 < \alpha < 1)$, *or* C^ω, *respectively. Suppose that* $S = \{X(w) \in \mathbb{R}^3 : w \in B\}$ *satisfies assumption* (A1) *as well as the following variational property* (A2):

S is stationary for the area functional (Dirichlet integral) in the class of all disk-type surfaces with boundary on T which partition $\mathscr{K}$ *into two parts of measures;* σ *meas* $\mathscr{K}$ *and* $(1 - \sigma)$meas $\mathscr{K}$.

Then $X(w)$ *is a real-analytic surface of constant mean curvature H on B. Moreover,* $X(w)$ *is of class* $C^{1,\beta}(\bar{B}, \mathbb{R}^3)$ *for every* $\beta \in (0, 1)$, *or of class* $C^{m,\alpha}(\bar{B}, \mathbb{R}^3)$, *or real-analytic on* $\bar{B}$, *respectively, and* $\bar{S} = S \cup \Sigma$ *intersects T orthogonally in the points of* $\Sigma = X(\partial B)$. *The mapping* $X(w)$ *satisfies the equation*

$$\Delta X = 2H X_u \wedge X_v$$

as well as the conformality relations

$$|X_u| = |X_v|, \qquad X_u \cdot X_v = 0$$

in B, and it intersects T perpendicularly.

We still have to make assumption (A2) precise. To this end, we suppose that the surface $S = \{X(w) : w \in B\}$ satisfies assumption (A1) and, furthermore, that S does not meet an $\mathbb{R}^3$-neighborhood $U(p)$ of some point $p \in T$. Let $U_1(p)$ be some smaller neighborhood of p with $U_1(p) \subset\subset U(p)$. Then we shall construct a vector field $Q(x) = (Q^1(x), Q^2(x), Q^3(x))$ on $\mathbb{R}^3$ such that the following holds:

(i) Q is of class C^1 on $\mathbb{R}^3$ and satisfies div $Q = 1$ in $\mathscr{K}$.
(ii) $Q|_T$ is a tangential vector field on $T - U_1(p)$; that is, the normal component $Q_n = n \cdot Q$ of Q satisfies $Q_n(x) = 0$ for all $x \in T - U_1(p)$. (Here n denotes the exterior normal of $T = \partial \mathscr{K}$.)

Such a vector field Q can be obtained in the special form $Q(x) = \operatorname{grad} f(x)$, where the scalar function $f(x)$ satisfies

$$\Delta f = 1 \quad \text{in } \mathscr{K}$$

$$\frac{\partial f}{\partial n} + \gamma f = 0 \quad \text{on } T = \partial \mathscr{K},$$

and γ stands for some sufficiently regular function on T which satisfies $\gamma \geqslant 0$, $\gamma \not\equiv 0$, and finally $\gamma \equiv 0$ on $T - U_1(p)$.

Let now $S = \{X = X(w) : w \in B\}$ be a solution of the partition problem. Then we infer from (i) and (ii) that, for the component Ω_1 of int$\{\mathscr{K} - S\}$ which appears in (A1), we have

$$\operatorname{meas} \Omega_1 = \iiint_{\Omega_1} dx^1 dx^2 dx^3 = \iiint_{\Omega_1} \operatorname{div} Q \, dx^1 dx^2 dx^3$$

$$= \iint_{\partial \Omega_1} Q_n \, dA = \iint_B Q(X) \cdot X_u \wedge X_v \, du \, dv.$$

That is,

$$\iint_B Q(X) \cdot X_u \wedge X_v \, du \, dv = \text{meas } \Omega_1. \tag{9}$$

The precise formulation of (A2) then reads as follows:

Assumption (A2). The surface $X(w)$ is stationary for the Dirichlet integral $D(Z) = \frac{1}{2}\iint_B |\nabla Z|^2 \, du \, dv$ within the class $\mathscr{C}^*$ of surfaces $Z \in \mathscr{C}(T)$ which satisfy

$$V^Q(Z) := \iint_B Q(Z) \cdot Z_u \wedge Z_v \, du \, dv = \sigma \cdot \text{meas } \mathscr{K} =: c. \tag{10}$$

Here, the set $\mathscr{C}(T)$ is defined by

$$\mathscr{C}(T) = \{Z : Z \in H_2^1(B, \mathbb{R}^3), Z(\partial B) \subset T\},$$

and $X(w)$ is called stationary within $\mathscr{C}^*$ if

$$\lim_{t \to 0} \frac{1}{t} \{D(X_t) - D(X)\} = 0$$

holds for every variation X_t of X which satisfies

$$V^Q(X_t) = c$$

and is of one of the two following types:

Type 1: The surfaces X_t, $|t| < t_0$, are of the form $X_t = X \circ \tau_t$, where $\{\tau_t\}_{|t| < t_0}$ is a family of diffeomorphisms from $\bar{B}$ to itself such that τ_0 is the identity, and that $\tau(t, w) := \tau_t(w)$ is of class C^1 on $(-t_0, t_0) \times \bar{B}$.

Type 2: The surfaces X_t are of the form

$$X_t(w) = X(w) + t\Psi(w, t), \qquad |t| < t_0,$$

where

$$D(\Psi(\cdot, t)) \leqslant C \qquad \text{for } |t| < t_0$$

with a bound C independent of t, and

$$\lim_{t \to 0} \Psi(w, t) = \Phi(w) \qquad \text{a.e. on } B$$

for some $\Phi \in H_2^1(B, \mathbb{R}^3)$.

The gist of this approach is that we have reduced the partition problem $\mathscr{P}$ to a free boundary problem for the Dirichlet integral with the subsidiary condition $V^Q(X) = c$. Using ideas from [18], one can establish the existence of a Lagrange parameter μ, and thus $\mathscr{P}$ is in fact reduced to a free boundary problem for the functional $D(X) + \mu V^Q(X)$ in the class $\mathscr{C}(T)$, without any further subsidiary condition.

The boundary regularity for solutions of this problem can, however, be proved by a suitable extension of the methods developed in [8]. For details we refer the reader to [9].

References

[1] H.W. Alt, *Die Existenz einer Minimalfläche mit freiem Rand vorgeschriebener Länge*, Arch. Rat. Mech. Anal. **51** (1973), 304–320.

[2] U. Dierkes, S. Hildebrandt, and H. Lewy, *On the analyticity of minimal surfaces at movable boundaries of prescribed length*, Preprint no. 753, SFB 72, Bonn (1985/86).

[3] G. Dziuk, *Über die Stetigkeit teilweise freier Minimalflächen*, manuscripta math. **36** (1981), 241–251.

[4] G. Dziuk, *On the length of the free boundary of a minimal surface*, Preprint SFB 72, Bonn (1984).

[5] E. Giusti, *The equilibrium configuration of liquid drops*, J. Reine Angew. Math. **321** (1981), 53–63.

[6] E. Gonzalez, U. Massari, and I. Tamanini, *On the regularity of boundaries of sets minimizing perimeter with a volume constraint*, Indiana Univ. Math. J. **32**, (1983), 25–37.

[7] M. Grüter, *Über die Regularität schwacher Lösungen des Systems* $\Delta x = 2H(x)x_u \wedge x_v$, Dissertation, Düsseldorf (1979).

[8] M. Grüter, *Regularity of weak H-surfaces*, J. Reine Angew. Math. **329** (1981), 1–15.

[9] M. Grüter, S. Hildebrandt, and J.C.C. Nitsche, *On the boundary behaviour of minimal surfaces with a free boundary which are not minima of the area*, manuscripta math. **35** (1981), 387–410.

[10] M. Grüter, S. Hildebrandt, and J.C.C. Nitsche, *Regularity for surfaces of constant mean curvature with free boundaries*, Acta. Math. **156** (1986), 119–152.

[11] M. Grüter, and J. Jost, (a) *On embedded minimal disks in convex bodies*, (b) *Allard-type regularity results for varifolds with free boundaries*, preprints (1984).

[12] S. Hildebrandt, *Minimal surfaces with free boundaries*, Miniconference on P.D.E., Canberra, C.M.A., A.N.U., August 1985, preprint.

[13] S. Hildebrandt and W. Jäger, *On the regularity of surfaces with prescribed mean curvature at a free boundary*, Math. Z. **118** (1970), 289–308.

[14] S. Hildebrandt and J.C.C. Nitsche, *Minimal surfaces with free boundaries*, Acta Math. **143** (1979), 251–272.

[15] S. Hildebrandt and J.C.C. Nitsche, *Optimal boundary regularity for minimal surfaces with a free boundary*, manuscripta math. **33** (1981), 357–364.

[16] S. Hildebrandt and J.C.C. Nitsche, *A uniqueness theorem for surfaces of least area with partially free boundaries on obstacles*, Archive Rat. Mech. Anal. **79** (1982), 189–218.

[17] S. Hildebrandt and J.C.C. Nitsche, *Geometric properties of minimal surfaces with free boundaries*, Math. Z. **184** (1983), 497–509.

[18] S. Hildebrandt and H.C. Wente, *Variational problems with obstacles and a volume constraint*, Math. Z. **135** (1973), 55–68.

[19] J. Jost, *Existence results for embedded minimal surfaces of controlled topological type*, Part II, SFB 72 Bonn, preprint no. 726 (1985).

[20] A. Küster, *An optimal estimate of the free boundary of a minimal surface*, J. Reine Angew. Math. **349** (1984), 55–62.

[21] H. Lewy, *On minimal surfaces with partially free boundary*, Comm. Pure Appl. Math. **4** (1951), 1–13.

[22] J.C.C. Nitsche, *The regularity of minimal surfaces on the movable parts of their boundaries*, Indiana Univ. Math. J. **21** (1971), 505–513.

[23] J.C.C. Nitsche, *Vorlesungen über Minimalflächen*, Springer-Verlag, Berlin-Heidelberg-New York (1975).

[24] J.C.C. Nitsche, *Stationary partitioning of convex bodies*, Arch. Rat. Mech. Anal. **89** (1985), 1–19.

[25] M. Struwe, *On a free boundary problem for minimal surfaces*, Invent. Math. **75** (1984), 547–560.

[26] R.-G. Ye, *A priori estimates for minimal surfaces with free boundary, which are not minima of the area*, preprint, Bonn, 1985.

Free Boundary Problems for Surfaces of Constant Mean Curvature

Michael Struwe

This survey describes a new existence result [21] for (disk-type) surfaces of prescribed constant mean curvature with free boundaries, and relates this result to some other well-known variational problems arising in differential geometry.

Rather than give complete mathematical proofs in this paper, we try to expose the underlying concepts in some current research on free boundary problems and to focus on aspects of common interest to both the pure and the more applied mathematician. For details we refer to [21].

Problem and Results

Let S be a compact surface in $\mathbb{R}^3$, and let $H \in \mathbb{R}$. We want to find surfaces of constant mean curvature H (for short "H-surfaces") of the type of the disk

$$B = \{w = (u,v) \in \mathbb{R}^2 \mid u^2 + v^2 < 1\}$$

with boundary on S and meeting S orthogonally along their boundary (like soap bubbles on the surface S).

If we introduce conformal coordinates $X = X(u,v)$ on such a surface, there results the following set of partial differential equations:

$$\Delta X = 2H\, X_u \wedge X_v \qquad \text{(parametric H-surface equation)} \qquad (1)$$

$$|X_u|^2 - |X_v|^2 = 0 = X_u \cdot X_v \qquad \text{(conformality relations)} \qquad (2)$$

$$X(\partial B) \subset S \qquad\qquad\qquad\qquad\qquad\qquad\qquad\qquad\qquad (3)$$

$$\qquad\qquad\qquad\qquad\qquad \text{(free boundary conditions)}$$

$$\partial_n X(w) \perp T_{X(w)}S, \ \forall w \in \partial B \qquad\qquad\qquad\qquad\qquad\qquad (4)$$

where, e.g., $X_u = (\partial/\partial u)X$, "$\wedge$" is the exterior product in $\mathbb{R}^3$, n is the outward unit normal on ∂B, "$\perp$" means orthogonal, and $T_P S$ denotes the tangent space to S at P.

Conversely, a solution of (1)–(4) will parametrize a surface of constant mean curvature H (away from branch points where $\nabla X(w) = 0$) that meets S orthogonally along ∂B. However, equations (1)–(4) constitute only a set of necessary conditions, and the parametric problem (1)–(4) may possess solutions with branch points, self-intersections, or intersections with S in the interior of their parameter domain.[1] Such surfaces would be physically unstable and hence will not be observed in soap bubble experiments.

Thus our parametric problem may have an even richer structure than we see in the physical world. The results and methods that we describe in this note may be a first approach towards a complete understanding of this structure—and hence in particular of the physics of soap bubbles.

Theorem 1. *Suppose $S \subset B_L(0) \subset \mathbb{R}^3$ is C^4-diffeomorphic to the standard sphere S^2 in $\mathbb{R}^3$. Then there exists a set $\mathcal{H} \subset \mathbb{R}$ containing $H = 0$ and lying dense in the interval $[-\frac{1}{L}, \frac{1}{L}]$ such that for any $H \in \mathcal{H}$ problem (1)–(4) has a nonconstant, regular solution whose image is contained in $B_L(0)$.*

Observe that, of course, any constant vector $X \equiv X_0 \in S$ trivially solves (1)–(4) for any $H \in \mathbb{R}$. Moreover, if S bounds a strictly convex region in $\mathbb{R}^3$ ("S is strictly convex"), any nonconstant minimal surface supported by S (a solution of (1)–(4) for $H = 0$) will be strictly unstable as a stationary point of Dirichlet's integral

$$D(X) = \frac{1}{2}\int_B |\nabla X|^2 \, dw.$$

We take this as justification of our simply referring to nonconstant solutions to (1)–(4) as "unstable H-surfaces on S".

Let us try to interpret Theorem 1 by looking at some related variational problems.

The Partition Problem

This is the "real" physical problem governing the shape of our soap bubbles:

"Given a surface S enclosing a volume vol(S) and a number $0 < V < $ vol(S), find a surface of least area "inside" S dividing the "interior" of S into two regions, one of which has volume V."

It is easy to verify that whenever such a minimal separating surface exists and is regular, it will meet S orthogonally and will have constant mean curvature H, the number H arising as a Lagrange multiplier associated with the (regular)

---

[1] Multiply connected minimal surfaces with free boundaries were studied, e.g., by Tolksdorf [23] and Jost [8].

volume constraint. (For the regularity question cf. the notes of I. Tamanini and S. Hildebrandt at this meeting.)

The above "partition problem" can be solved by using methods of geometric measure theory. Cf. U. Massari [10] and J. Taylor [22]. Simple examples, however, will convince the reader that the least area solutions obtained in this manner may have a very complicated topological type. Stationary partitioning hypersurfaces show an even higher variability, as is amply illustrated by examples given by J.C.C. Nitsche [11]. In contrast, the existence of partitioning surfaces of a prescribed topological type is largely open.

Another obviously related problem is the following.

The Plateau Problem for H-Surfaces

In the Plateau problem for H-surfaces, instead of the free boundary conditions (3), (4) the following condition is imposed:

$$X|_{\partial B}: \partial B \to \Gamma \text{ is a parametrization of a given Jordan curve } \Gamma \subset \mathbb{R}^3. \quad (5)$$

For $H = 0$ equations (1), (2), (5) of course constitute the classical Plateau problem for minimal surfaces solved by J. Douglas and T. Radó in 1930–31. Generalizations for $H \neq 0$ were obtained beginning in the 1950s under various geometric restrictions on H. Finally, in 1970, S. Hildebrandt [6] obtained stable solutions to (1), (2), (5) under the condition that

$$|H|\,\|\Gamma\|_{L^\infty} < 1,$$

a condition which is best possible for the circle.

At first glance there is a striking similarity between this result and our Theorem 1 for the free boundary problem. Note, however, that in the case of free boundaries, interesting solutions are in general unstable while stable solutions trivially exist for all $H \in \mathbb{R}$.

For the Plateau problem unstable solutions (for "small" $H \neq 0$) were independently obtained by Brezis–Coron [1] and the author [17]—with an important contribution by Steffen [16]. These results were extended in [19] where the following result was established: Whenever there is a stable solution to the Plateau problem (1), (2), (5) for $H \neq 0$, there will always also be an unstable solution.

By analogy, in the free boundary problem we would expect unstable solutions to exist for any $H \neq 0$ on any (compact) supporting surface S—regardless of its topological type. However, technical difficulties essentially due to the presence of complete H-surfaces (spheres) inside S for large H considerably complicate the proofs.

Some Historical Remarks

Partition and free boundary problems for minimal surfaces have a long history, going back to the famous problem posed by Gergonne in 1816 to find a surface of least area which meets a pair of diagonals lying on opposite faces of the cube and crossed at right angles.

Together with the least area solution to Gergonne's problem, H.A. Schwarz in 1872 also presented a countable family of stationary minimal surfaces inside the cube and satisfying the required boundary conditions [14]. Schwarz then proceeded to propose generalizations of Gergonne's boundary condition (Schwarz' chains) which naturally led to considering minimal surfaces with free boundaries on manifolds S.

If S is compact and genus $(S) > 0$, R. Courant [2, p. 199 ff.] in the 1940s obtained minimal surfaces on S by means of a variant of Dirichlet's principle: Let Γ be a contour which is not contractible in $\mathbb{R}^3 \backslash S$; then minimize Dirichlet's integral among surfaces whose boundaries $X|_{\partial B}$ link with Γ!

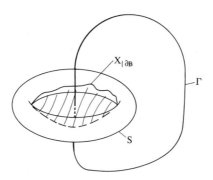

For genus $(S) = 0$ this idea cannot be applied. In fact, we had seen earlier that for strictly convex S only the trivial (constant) solutions to (1)–(4) for $H = 0$ are stable and can be obtained as relative minima of D. This case thus remained open until 1983 when tools developed for harmonic mappings of Riemannian surfaces (cf. Sacks–Uhlenbeck [13]) permitted to state the following existence result obtained in [18].

Theorem 2. *Suppose $S \subset \mathbb{R}^3$ is C^4-diffeomorphic to S^2. Then there exists a non-constant solution to* (1)–(4) *for $H = 0$.*

About the same time, B. Smyth [15] presented existence results for minimal surfaces inside a tetrahedron. However, his proof is strictly algebraic and works only for this case.

Later, Theorem 2 was improved for convex (or more generally H-convex) surfaces, and even the existence of embedded minimal discs inside such supporting surfaces was established (cf. Grüter–Jost [5], Jost [8]). (Recall that a surface S is H-convex if S has non-negative mean curvature with respect to the interior normal.) To motivate these existence results and to illustrate how harmonic mappings should be helpful in solving (1)–(4), let us introduce

The Variational Problem.

For ease of exposition we restrict ourselves to the case $H = 0$. Then solutions to (1)–(4) can be characterized as the "critical points" of Dirichlet's integral on the space

$$C(S) = \{X \in H^{1,2}(B; \mathbb{R}^3) | X(\partial B) \subset S\}$$

of maps X with square integrable distributional derivative and boundary on S. By (1) we may immediately restrict ourselves to the subspace

$$C_0(S) = \{X \in C(S) | \Delta X = 0\}$$

of surfaces whose coordinate functions are harmonic. Note that $H^{1,2}(B) \to L^2(\partial B)$ by restriction $X \to X|_{\partial B}$, and X is a.e. defined on ∂B.

By harmonic extension we may consider the space of closed curves

$$M = \{x \in H^{1,2}(\partial B; \mathbb{R}^3) | x(\partial B) \subset S\}$$

as a dense subspace of $C_0(S)$. These curves have slightly better regularity properties than the restriction $X|_{\partial B}$ of $X \in C_0(S)$ in general would have. This causes M to be a Hilbert manifold while $C_0(S)$ is not.

If we now consider D on M we note an analogy between our variational problem and the following problem.

Closed Geodesics on S

In abstract terms, closed geodesics on S correspond to the non-constant critical points of a certain functional on M which is equivariant with respect to the 0(2)-action

$$\phi \to \phi + \phi_0 \quad \text{(rotation by an angle } \phi_0\text{)}$$

$$\phi \to 2\pi - \phi \quad \text{(reflection)}. \tag{6}$$

Here, of course, ϕ denotes the angular coordinate on ∂B. The following theorem is essentially due to Ljusternik and Schnirelman. Introducing the Palais–Smale condition (cf. [12]) allows us to state a streamlined version of their classical result [9].

Theorem 3. *Let $E \in C^1(M)$ be equivariant under the 0(2)-action (6). Moreover, assume that $E(x) \geq 0$, $E(x) = 0$ iff $x \equiv$ const. Also, let the Palais–Smale condition be satisfied:*

> *Any sequence $\{x_m\} \subset M$ where $E(x_m) \leq c$ uniformly while $dE(x_m) \to 0$, is relatively compact.* (C)

Then E possesses at least 3 non-constant critical points which are distinct modulo (6).

With Theorem 3 available, existence results for unstable minimal surfaces on $S \cong S^2$ would be easy to obtain. Unfortunately, our functional D is equivariant with respect to a much larger group: The conformal group of the disc B. I.e. there holds

$$D(X \circ g) = D(X) \tag{7}$$

for any $X \in H^{1,2}(B; \mathbb{R}^3)$ and any conformal diffeomorphism $g: \bar{B} \to \bar{B}$. Moreover, if $X \not\equiv \text{const.}$, the conformal orbit of X: $\{X \circ g | g: \bar{B} \to \bar{B} \text{ is conformal}\}$ is not relatively compact in $H^{1,2}$. In particular, letting X be the non-constant minimal surface guaranteed by Theorem 2, the conformal orbit of X—which by (7) consists of critical points of D—is not relatively compact, and the Palais–Smale condition (C) cannot hold for D on $C_0(S)$ or any space where this group of symmetries is acting.

Still it is generally believed that any (smooth) surface $S \cong S^2$ will support at least three distinct non-constant minimal surfaces; and some partial results exist that strongly support this conjecture [8], [15]. More generally, we believe that there are at least three distinct non-constant H-surfaces on S lying inside $B_L(0)$, provided $S \subset B_L(0)$, $S \cong S^2$, $|H| < 1/L$.

Help to assure the existence of at least one non-constant solution to (1)–(4) comes from a different direction.

Harmonic Maps from S^2 to S

To illustrate the relation of harmonic maps $S^2 \to S$ with our problem, let us argue heuristically as follows: Let $\delta > 0$ and suppose X is a minimal surface supported by S whose image lies in a δ-neighborhood of S. By compactness and regularity of S, for sufficiently small $\delta > 0$ any point $X(w)$ on X has a unique projection $\hat{X}(w)$ onto S, where

$$|\hat{X}(w) - X(w)| = \inf_{Y \in S} |Y - X(w)|,$$

and a unique reflection image $\tilde{X}(w) = 2\hat{X}(w) - X(w)$.

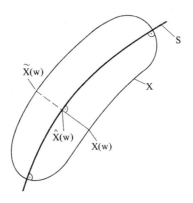

Since reflection is involutory, from harmonicity of X we derive that

$$\Delta \tilde{X} = 2\tilde{\Gamma}(\tilde{X})(\nabla \tilde{X}, \nabla \tilde{X}) \quad \text{in } B$$

with a bounded bilinear form $\tilde{\Gamma}$ whose coefficients smoothly depend on $\tilde{X}$. Equations (3)–(4) moreover give the boundary condition

$$\partial_n \tilde{X} = -\partial_n X \quad \text{on } \partial B.$$

Hence, for $\hat{X} = \frac{1}{2}(X + \tilde{X})$ we obtain the equation

$$\Delta \hat{X} = \tilde{\Gamma}(\tilde{X})(\nabla \tilde{X}, \nabla \tilde{X}) \quad \text{in } B \tag{8}$$

together with the homogeneous Neumann boundary condition

$$\partial_n \hat{X} = 0 \quad \text{on } \partial B. \tag{9}$$

In the limit $\delta \to 0$, $X = \hat{X} = \tilde{X}$ coincide and $\tilde{\Gamma}(\tilde{X}) = \Gamma(X)$, where the coefficients of the bilinear form Γ are the Christoffel symbols of the second fundamental form of S. Thus (8) in the limit $\delta \to 0$ becomes the equation for harmonic mappings $\hat{X}: B \to S$. By (9) we may "double" $\hat{X}$

$$\hat{X}(w) = \begin{cases} \hat{X}(w), & w \in B \\ \hat{X}\left(\dfrac{w}{|w|^2}\right), & w \notin B \end{cases}$$

to obtain a harmonic map $\hat{X}: \mathbb{R}^2 \to S$. Now observe that $\mathbb{R}^2$ is conformally equivalent to $S^2 - \{p\}$. Hence, $\hat{X}$ induces a harmonic map $\hat{X}: S^2 - \{p\} \to S$. Finally, by a result of Sacks and Uhlenbeck [13, Theorem 3.6], $\hat{X}$ may be smoothly extended to a harmonic map $\hat{X}: S^2 \to S$.

This argument at least formally explains the deep connections between our problem and harmonic maps of compact surfaces—relations which of course are well-known to the experts and have already been used to derive regularity results for minimal surfaces with free boundaries, cf., e.g. [7]. In order to prove the existence of unstable minimal surfaces we are thus tempted to try the following.

The Sacks–Uhlenbeck Approximation Method

For our problem this means approximating $C(S)$ by manifolds

$$C_\alpha(S) = C(S) \cap H^{1,2\alpha}(B; \mathbb{R}^3), \quad \alpha > 1,$$

and replacing D on $C_\alpha(S)$ by the functional

$$D_\alpha(X) = \frac{1}{2} \int_B [(1 + |\nabla X|^2)^\alpha - 1] \, dw, \quad \alpha > 1.$$

D_α for $\alpha > 1$ again is $0(2)$-equivariant (but not conformally invariant) and satisfies the Palais–Smale condition. By Theorem 3 there exist at least 3 distinct critical points X_α of D_α on $C_\alpha(S)$ for any $\alpha > 1$ (with uniform bounds on the critical values). As $\alpha \to 1$, finally, either a subsequence $X_\alpha \to X$ in $C(S)$ and X is a non-constant solution to (1)–(4) (for $H = 0$), or we may encounter the phenomenon of "separation of spheres": The Dirichlet integrals of X_α concentrate at points in $\bar{B}$ where singularities develop. But these singularities only form in the parametrizations of X_α—not in the surfaces themselves. Hence, if we rescale suitably around these points of concentration in the limit $\alpha \to 1$ we find a smooth surface $\bar{X}$—which turns out to be a minimal surface supported by S, parametrized over a half-plane $\mathbb{R}^2_+$. By conformal equivalence $\mathbb{R}^2_+ \cong B$ and the regularity results [3], [4], $\bar{X}$ induces a non-constant, regular solution to (1)–(4) for $H = 0$; cf. [18] for details.

This proves Theorem 2. Extensions to the case $H \neq 0$ require certain a priori estimates on H-surfaces which are not known even in the case of the Plateau problem, and a different approach is needed.

The Evolution Problem

Again our inspiration comes from harmonic maps. In 1984 a direct approach to (unstable) harmonic maps of Riemannian surfaces was developed based on the analysis of the time-dependent problem [20]. The counterpart of this evolution problem for H-surfaces on S takes the form

$$\partial_t X - \Delta X + 2H X_u \wedge X_v = 0 \qquad \text{on } B \times \,]0, T[, \tag{10}$$

$$X(\partial B \times [0, T[) \subset S, \tag{11}$$

$$\partial_n X(w, t) \perp T_{X(w,t)} S, \qquad \forall w \in \partial B, \, t \in \,]0, T[, \tag{12}$$

$$X|_{t=0} = X_0 \in C(S) \tag{13}$$

on a cylinder $B \times [0, T[$. By reflection in S a solution to (10)–(13) may be extended to a solution of the evolution equation for H-surfaces in a Riemannian metric on a larger cylinder. The estimates of [20] thus convey to our case and, in particular, for $H = 0$ we obtain the following theorem, cf. [21, Theorem 3.2].

Theorem 4. For any $X_0 \in C(S)$ there exists a global distribution solution X to (10)–(13) which is defined and regular on $\bar{B} \times \,]0, \infty]$ with exception of at most finitely many points $(w^{(k)}, T^{(k)})$, $T^{(k)} \leq \infty$, $1 \leq k \leq K$.

At a singularity $(w^{(k)}, T^{(k)})$ a non-constant minimal surface on S separates in the sense that for suitable sequences $t_m^{(k)} \nearrow T^{(k)}$, $w_m^{(k)} \to w^{(k)}$, $R_m^{(k)} \searrow 0$ the rescaled surfaces

$$X(w_m^{(k)} + R_m^{(k)} w, t_m^{(k)}) \xrightarrow[(m \to \infty)]{} \bar{X}^{(k)} \qquad \text{in } H^{2,2}_{loc}(\mathbb{R}^2_+; \mathbb{R}^3),$$

where $\bar{X}^{(k)}$ is conformal to a non-constant regular solution to (1)–(4) for $H = 0$.

If $T^{(k)} < \infty$ for all k, then as $t \to \infty$

$$X(\cdot, t) \to \bar{X} \qquad \text{in } H^{2,2}(B; \mathbb{R}^3),$$

and $\bar{X}$ solves (1)–(4).

With Theorem 4 at our disposal we can now very easily re-prove Theorem 2: Either for some initial value $X_0 \in C(S)$ the flow (10)–(13) through X_0 (for $H = 0$) becomes singular at a point $(\bar{w}, \bar{T})$, and Theorem 4 guarantees the existence of a non-constant minimal surface on S. Or the flow (10)–(13) is globally regular for all initial data and continuously retracts $C(S)$ onto the set of critical points of D. But in this event the Palais–Smale condition is no longer needed for the assertion of Theorem 3, and we can even find 3 distinct minimal surfaces on S! (However, one must exclude multiple coverings.) A way to higher multiplicity results for unstable minimal surfaces thus would be to exclude singularities for the flow (10)–(13).

The above method extends to $H \neq 0$, $|H| < 1/L$, but becomes more involved. The possible "gaps" in the set $\mathscr{H}$ of admissible curvatures correspond to possible points of discontinuity of a certain monotone function. They also reflect the absence of "reasonable" a priori bounds for H-surfaces—and indicate another direction of further research.

A Word on Modelling

The interplay between cohesive (surface tension) and adhesive (boundary) forces may result in a variety of different models for the motions of soap films with free boundaries on a supporting surface S. To take a look at two extreme cases let us consider a plane disk moving upwards in a conical segment of S. If surface tension dominates, the film will move as a planar surface under the pull of the boundary forces, i.e. governed by the system

$$\Delta X = 0, \tag{14}$$

$$\partial_t X = -\partial_n X'', \tag{15}$$

where $\partial_n X''$ indicates the component of $\partial_n X$ tangent to S, while predominance of adhesive forces will immediately lead to the orthogonality condition (3)—corresponding to our model (10)–(13).

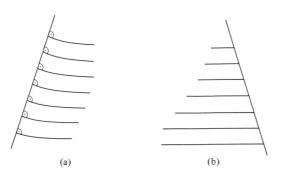

(a)   (b)

Soap film motion modelled by (10)–(13)—(a)—compared with motion modelled by (14), (15)—(b)

Mathematically, (10)–(13) has proved much simpler to handle than (14), (15) where, e.g., uniqueness is hard to obtain. Equations (10)–(13) can also be realized much easier physically: Changing the contact angle to 90° reduces potential and requires arbitrarily little kinetic energy.

However, from the soap film experiment we may not be able to decide between (10)–(13), resp. (14), (15): When we look at a contact point of a soap film with the container wall, all we see is a lump of soap.

Acknowledgments. These notes are based on a lecture given at the "Symposium on variational methods for free surface interfaces" at Menlo Park, Cal., Sept. 13, 1985. It is my pleasure to thank Robert Finn and Paul Concus for the careful organization of this meeting. I also thank the AMS and the SFB 72 at the University of Bonn for their generous support.

References

1. H. Brezis and J.-M. Coron, *Multiple solutions of H-systems and Rellich's conjecture*, Comm. Pure Appl. Math. **37** (1984), 149–187.
2. R. Courant, *Dirichlet's principle, conformal mapping and minimal surfaces*, Interscience, New York, 1950.
3. G. Dziuk, C^2-*regularity for partially free minimal surfaces*, Math. Z. **189** (1985), 71–79.
4. M. Grüter, S. Hildebrandt, and J.C.C. Nitsche, *Regularity for stationary surfaces of constant mean curvature with free boundaries*, Acta Math. (in press).
5. M. Grüter and J. Jost, *On embedded minimal discs in convex bodies* (Preprint).
6. S. Hildebrandt, *On the Plateau problem for surfaces of constant mean curvature*, Comm. Pure Appl. Math. **23** (1970), 97–114.
7. S. Hildebrandt and J.C.C. Nitsche, *Minimal surfaces with free boundaries*, Acta Math. **143** (1979), 251–272.
8. J. Jost, *Existence results for embedded minimal surfaces of controlled topological type* (Preprint).
9. L. Ljusternik and L. Schnirelman, *Existence de trois géodésiques fermées sur toute surface de genre o*, C.R. Acad. Sci. Paris **188** (1929), 534–536.
10. U. Massari, *Esistenza e regolaritá delle ipersuperfici di curvatura media assegnata in $\mathbb{R}^n$*, Arch. Rat. Mech. Anal. **55** (1974), 357–382.
11. J.C.C. Nitsche, *Stationary partitioning of convex bodies*, Arch. Rat. Mech. Anal. **89** (1985), 1–19.
12. R.S. Palais and S. Smale, *A generalized Morse theory*, Bull. AMS **70** (1964), 165–171.
13. J. Sacks and K. Uhlenbeck, *The existence of minimal immersions of 2-spheres*, Ann. Math. **113** (1981), 1–24.
14. H.A. Schwarz, *Gesammelte Mathematische Abhandlungen*, Band I, Springer, Berlin, 1890.
15. B. Smyth, *Stationary minimal surfaces with boundary on a simplex*, Inv. Math. **76** (1984), 411–420.
16. K. Steffen, *On the nonuniqueness of surfaces with prescribed constant mean curvature spanning a given contour*, Arch. Rat. Mech. Anal. **94** (1986), 101–122.
17. M. Struwe, *Nonuniqueness in the Plateau problem for surfaces of constant mean curvature*, Arch. Rat. Mech. Anal. **93** (1986), 135–157.

18. M. Struwe, *On a free boundary problem for minimal surfaces*, Inv. Math. **75** (1984), 547–560.
19. M. Struwe, *Large H-surfaces via the mountain-pass-lemma*, Math. Ann. **270** (1985), 441–459.
20. M. Struwe, *On the evolution of harmonic mappings of Riemannian surfaces*, Comm. Math. Helv. **60** (1985), 558–581.
21. M. Struwe, *The existence of surfaces of constant mean curvature with free boundaries* (Preprint).
22. J.E. Taylor, *Boundary regularity for solutions to various capillarity and free boundary problems*, Comm. PDE **2** (1977), 323–357.
23. P. Tolksdorf, *A parametric variational principle for minimal surfaces of varying topological type*, J. Reine Angew. Math. **354** (1984), 16–49.

On the Existence of Embedded Minimal Surfaces of Higher Genus with Free Boundaries in Riemannian Manifolds

Jürgen Jost

0. Introduction

In this chapter we consider the following configuration: a Riemannian manifold X of bounded geometry, some closed Jordan curves Γ_j, and a supporting surface ∂K, disjoint from the Γ_j. We further assume that the Γ_j are contained in a suitable barrier ∂A of nonnegative mean curvature (cf. §2 for details).

We seek an embedded minimal surface of specified genus g, with fixed boundary curves Γ_j and (possibly) free boundary curves on the supporting surface ∂K.

We find such a surface, provided a suitable Douglas type criterion holds (which is a natural condition for such problems; cf. [6]). Our results can be considered as extensions of our corresponding ones for the genus O case in [7]. We can actually treat two cases: In Theorem 2 below, we do not specify the number of free boundary curves of our minimal surface, whereas in Theorem 3 we impose a curvature condition on ∂K, enabling us to fix this number a priori, again by a Douglas type condition.

Let us mention [2], where such existence question for embedded minimal surfaces were considered in the absence of a free boundary. Our method (which is different from [2]) can produce similar results, but actually the situation considered in the present paper is considerably more general than the one of [2]. We note, however, that we always need the assumption that the existence of an embedded surface with the specified topological data is topologically possible. Hence, we cannot deduce topological consequences from our results. The same seems to be true of [2], however, in contrast to the genus O case treated by Meeks–Yau in [11]. Of course, if we have no fixed boundary, then our limit surface can be a double covering of a non-orientable surface as in [2].

We shall combine three different strings of reasoning:

—Our investigation of manifolds with free boundaries in [7] that extends the interior regularity results of [1].
—Approximation arguments of Meeks–Yau ([11], [12]) as employed in [7].
—Our techniques of [6] for parametric minimal surfaces of higher genus based on conformal representation and hyperbolic geometry.

Actually, for our present purpose, we need to represent a compact surface of genus g with boundary conformally as a hyperbolic surface with an appropriate number of geodesic disks removed. Although, at first sight, one might believe that such a representation theorem is known, it was recently found out that even in the case of genus 1, no complete proof was available. Such a proof was then supplied by Strebel [16]. Strebel's result was used by Schüffler [17] when proving an index theorem for minimal surfaces of genus 1. We therefore supply a proof for genus ≥ 2, along lines rather different from the arguments of [16]. Of course, our proof can be modified to cover the cases of genus 1 and 0 (cf. [5] for the latter) as well, and of course in these cases the argument becomes considerably simpler.

The needed result on conformal representation is proved in §1. We shall use some of the basic concepts of Teichmüller theory. All geometric notions in §1 shall always refer to hyperbolic metrics on the surfaces considered, i.e. metrics of constant negative Gauss curvature. The existence results for embedded minimal surfaces shall be presented in §2. Some of the arguments can be taken over from [7], and we shall only sketch these.

This paper was stimulated by the conference on free boundary value problems in Vallombrosa, Sept. 1985, and I would like to thank Robert Finn for the pleasant atmosphere he created as the organizer of this conference. I also gratefully acknowledge support from the Stiftung Volkswagenwerk and MSRI, Berkeley, where this paper was written.

1. A Result on Conformal Representation of Compact Surfaces with Boundary

Theorem 1. *Suppose S is a compact oriented surface with boundary, homeomorphic to a closed surface of genus $p \geq 2$ with k disks removed. Suppose that S has a metric and can be covered by a finite number of coordinate charts so that in each chart the metric tensor is given by bounded measurable functions g_{ij} satisfying almost everywhere*

$$g_{11}g_{22} - g_{12}^2 \geq \lambda > 0, \tag{1.1}$$

where λ is a fixed constant. Then there exists a surface Σ_0 (compact, oriented, of genus p) with a hyperbolic metric (i.e. a quotient of the upper half plane $H = \{z = x + iy \in \mathbb{C}, y > 0\}$ by a discrete, fixed-point free group of hyperbolic motions), a surface Σ with boundary obtained by removing k geodesic disks from Σ_0, and a conformal homeomorphism

$$\tau : \Sigma \to S.$$

Conformality here means that in local coordinates almost everywhere

$$g_{ij}\frac{\partial \tau^i}{\partial x}\frac{\partial \tau^j}{\partial x} = g_{ij}\frac{\partial \tau^i}{\partial y}\frac{\partial \tau^j}{\partial y} \tag{1.2}$$

$$0 = g_{ij}\frac{\partial \tau^i}{\partial x}\frac{\partial \tau^j}{\partial y},$$

where $z = x + iy$ is a conformal coordinate on Σ. If S is of class $C^{k,\alpha}$, C^∞, or C^ω, then so is τ. In particular, if $S \in C^{1,\alpha}$, then (1.2) is satisfied everywhere, and τ is a diffeomorphism ($0 < \alpha < 1$).

Remark. Similar statements hold, of course, for $p = 0$ and $p = 1$, and they can be proved by easy modifications (actually simplifications) of the subsequent proof. The case $p = 0$ was treated in [5; Chapter 3], but was of course known before, whereas the case $p = 1$ was recently treated by Strebel [16] by a completely different method.

PROOF OF THEOREM 1. In [6], we demonstrated that the Schottky double $\bar{S}$ of S can be uniformized by a closed hyperbolic surface $\bar{\Sigma}$ with an isometric involution. The fixed-point set of this involution consists of k circles, corresponding to the k boundary curves of S (cf. Corollary 1.1 of [6]). The genus of $\bar{\Sigma}$ is $2p + k - 1$. We let the k boundary circles divide $\bar{\Sigma}$ into two surfaces Σ_1 and Σ_2 (each of which then is conformally equivalent to S). The boundary curves of Σ_1 and Σ_2 are closed geodesics of $\bar{\Sigma}$ (with respect to the hyperbolic metric).

It therefore suffices for our purpose to show that any such surface Σ_1 is conformally equivalent to a surface Σ as in the statement of the theorem, i.e. a hyperbolic surface with k geodesic disks removed.

We let $\bar{T}_{p,k}$ be the space of all equivalence classes of closed hyperbolic surfaces of genus $2p + k - 1$, possessing an isometric involution i the fixed-point set of which consists of k closed geodesics. In order to define an equivalence class, we choose an underlying topological model $\bar{\Sigma}'$ of our surfaces and consider two hyperbolic structures as equivalent if there exists a conformal (hence isometric) diffeomorphism between them which is homotopic to the identity of $\bar{\Sigma}'$.

$\bar{T}_{p,k}$ is diffeomorphic to $\mathbb{R}^{6p+3k-6}$. Actually, let S be any surface of genus $2p + k - 1$ with the required symmetry. $\bar{T}_{p,k}$ can then be identified with the space of those holomorphic quadratic differentials on S that respect this symmetry.

Likewise, we form the space $T_{p,k}$ of equivalence classes of surfaces Σ that can be obtained by removing k geodesic disks from a closed hyperbolic surface of genus p.

Let $R^1, R^2 \in T_{p,k}$ be obtained by removing from hyperbolic surfaces R_0^1, R_0^2 respectively, disks $D(p_j^i, r_j^i)$ ($p_j^i \in R_0^i$, $r_j^i \in (0, \infty)$, $j = 1, \ldots, k$) ($i = 1, 2$ resp.). Here, of course, $D(p, r)$ is the geodesic disk with center p and radius r. Suppose there exists a conformal diffeomorphism h between R^1 and R^2 mapping $\partial D(p_j^1, r_j^1)$ onto $\partial D(p_j^2, r_j^2)$ for each $j = 1, \ldots, k$. We then claim that h extends to a conformal diffeomorphism (hence isometry) $h_0: R_0^1 \to R_0^2$. It then follows that if h is homotopic to the identity of the underlying topological surface, then R_0^1 and R_0^2 belong to the same equivalence class, i.e. can be identified. Hence $p_j^1 = p_j^2, r_j^1 = r_j^2$ for all $j = 1, \ldots, k$. Hence the hyperbolic structures on R^1 and R^2 agree, and all the centers and radii of the resp. removed disks agree. To show the claim, we let $|\partial h|$ be the norm of the derivative of h (with respect to the hyperbolic metrics). We recall the well-known formula (cf. e.g. [5; (12.2.2)])

$$\Delta \log |\partial h|^2 = -1 + |\partial h|^2.$$

It follows from the maximum principle that at an interior maximum z_1 of $|\partial h|$, we have

$$|\partial h|(z_1) \leqslant 1,$$

and that at an interior minimum z_2 of $|\partial h|$, we have

$$|\partial h|(z_2) \geqslant 1.$$

On the other hand, since the boundary curves of both R^1 and R^2 are circles, it follows from the Schwarz reflection principle that h can be continued across the boundary circles (at least locally into a full neighborhood of each boundary circle). Also, if $z \mapsto \tilde{z}$ denotes the reflection, $|\partial h|(z) = |\partial h|(\tilde{z})$. Therefore, the same reasoning also applies to boundary maxima resp. minima of $|\partial h|$, and since $|\partial h|$ has to assume its maximum and minimum on the compact surfaces R^1 and R^2 (with boundary), we conclude

$$|\partial h| \equiv 1.$$

Hence h is an isometry. In particular, the mapping of the boundary circles is an isometry, and hence h can be trivially continued to an isometry $h_0: R^1 \to R^2$.

We draw some consequences from this observation. Firstly, $T_{p,k}$ has a structure of a differentiable manifold, coordinates given by $6p - 6$ parameters for the Teichmüller space of surfaces of genus p plus $3k$ coordinates for the centers and radii of the removed geodesic disks. In particular, $\dim T_{p,k} = 6p + 3k - 6$. Actually, one can check right away that $T_{p,k}$ is diffeomorphic to $\mathbb{R}^{6p+3k-6}$. This will, however, also be a consequence of our subsequent reasoning.

We can form a quotient $\overline{M}_{p,k}$ of $\overline{T}_{p,k}$ (and similarly a quotient $M_{p,k}$ of $T_{p,k}$) by identifying hyperbolic structures between which there exists a conformal diffeomorphism (hence isometry), not necessarily homotopic to the identity this time.

By Corollary 1.1 of [6] again, every element of $M_{p,k}$ is conformally equivalent to an element of $\overline{M}_{p,k}$. This gives a canonical map

$$m: M_{p,k} \to \overline{M}_{p,k}.$$

This map is injective by the above arguments, and induces a differentiable map

$$t: T_{p,k} \to \overline{T}_{p,k}.$$

We shall show that m is a proper map. Since $T_{p,k}$ and $\overline{T}_{p,k}$ have the same dimension, it will then follow that t is also surjective. In other words, every element of $\overline{T}_{p,k}$ is conformally equivalent to an element of $T_{p,k}$, thus finishing the proof by appealing to Corollary 1.1 of [6].

To show properness, we look at the following compact subsets of $\overline{M}_{p,k}$ and $M_{p,k}$. For $\varepsilon > 0$, let $\overline{M}^\varepsilon_{p,k}$ be the set of all hyperbolic structures in $\overline{M}_{p,k}$ for which the length of the shortest closed geodesic is at least ε. For $\delta > 0$, let $M^\delta_{p,k}$ be the set of all hyperbolic structures Σ in $M_{p,k}$ satisfying the following conditions:

(i) If Σ is obtained by removing disks $D(p_j, r_j)$ $(j = 1, \ldots, k)$ from a hyperbolic surface Σ_0, and if $i(p)$ denotes the injectivity radius of a point $p \in \Sigma_0$, then

$$r_j \geqslant \delta \quad \text{for } j = 1, \ldots, k$$

$$r_j \leqslant i(p_j) - \delta$$

$$r_{j_1} + r_{j_2} \leqslant \text{dist}(p_{j_1}, p_{j_2}) - \delta \quad \text{for } j_1 \neq j_2;$$

(ii) the length of the shortest closed geodesic in Σ_0 is at least δ.

That $\overline{M}^\varepsilon_{p,k}$ for $\varepsilon > 0$ is a compact subset of $\overline{M}_{p,k}$ follows from Mumford's compactness theorem ([13]). Both Mumford's theorem in the case of hyperbolic surfaces and the corresponding statement for $M^\delta_{p,k}$ can be proved by arguments of elementary hyperbolic geometry. If, e.g., $(\Sigma^n)_{n \in \mathbb{N}}$ is a sequence in $M^\delta_{p,k}$ $(\delta > 0)$, one represents every Σ^n by a fundamental region in H with appropriate disks removed and readily shows that a subsequence of the fundamental regions converges to the fundamental region of a limiting surface (again with appropriate disks removed). (A detailed proof along these lines will also be given in a forthcoming exposition of the author.)

In order to finish the proof, we therefore only need to verify that for every $\varepsilon > 0$, there is some $\delta > 0$ with

$$t^{-1}(\overline{M}^\varepsilon_{p,k}) \subset M^\delta_{p,k}.$$

Otherwise, there is some $\varepsilon > 0$, a sequence $(\delta_n)_{n \in \mathbb{N}}$, $\delta_n \to 0$, and hyperbolic structures

$$\Sigma^n \in M_{p,k} \setminus M^{\delta_n}_{p,k}$$

with

$$t(\Sigma^n) \subset \overline{M}^\varepsilon_{p,k}.$$

We shall show that this is impossible by showing that for any diffeomorphism

$$u^n: \Sigma^n \to \Sigma' \in \overline{M}^\varepsilon_{p,k}$$

the energy

$$E(u^n) = \frac{1}{2} \int_{\Sigma^n} |du^n|^2$$

tends to infinity as $n \to \infty$, thus contradicting the fact that the energy of a conformal diffeomorphism equals the area of the image which is $2\pi(6p + 3k - 6)$ in the present case.

The surfaces Σ^n can tend to degeneration in several ways, and hence several cases have to be considered. That the injectivity radius of Σ^n tends to zero can be ruled out using the collar lemma as in [14] or [6]. The argument in the other cases is similar. Let us, for example, treat the case where Σ^n is obtained from a hyperbolic surface Σ_0^n by removing a disk $D(p^n, r^n)$ with $r^n \to 0$ as $n \to \infty$. We then look at the annulus

$$A^n = \{z \in \Sigma_0^n : r^n \leqslant \mathrm{dist}(z, p^n) \leqslant \sqrt{r^n}\} \subset \Sigma^n.$$

This annulus is conformally equivalent to a cylinder with radius 1 and height $l_n \to \infty$ as $n \to \infty$, $C^n = \{(\varphi, z) : \varphi \in S^1, 0 \leqslant z \leqslant l_n\}$.

Since the energy of a map is invariant under conformal changes of the domain, we can look at u^n as a map from C^n into a surface contained in $\overline{M}_{p,k}^{\varepsilon}$. For each $z \in [0, l_n]$, the length of $u^n\{(\varphi, z) : \varphi \in S^1\}$ is at least ε. Hence

$$\varepsilon \leqslant \int_{\varphi=0}^{2\pi} |du_n|(\varphi, z)\, d\varphi \leqslant (2\pi)^{1/2} \left(\int_{\varphi=0}^{2\pi} |du_n|^2(z, \varphi)\, d\varphi \right)^{1/2}.$$

Integrating this inequality with respect to z, we obtain

$$\frac{\varepsilon^2 l_n}{2\pi} \leqslant \int_{z=0}^{l_n} \int_{\varphi=0}^{2\pi} |du_n|^2\, d\varphi\, dz \leqslant 2E(u^n),$$

hence $E(u^n) \to \infty$, as $n \to \infty$.

Regarding the other possible degenerations, we observe that the collar lemma argument excluding pinching of closed geodesics won't be affected by the possible presence of some of the disks $D(p_j^n, r_j^n)$ inside the collar, that the above argument also applies if a shrinking disk also becomes tangent to another one in the limit, and that we can look at a neighborhood of the shortest geodesic from $\partial D(p_{j_1}^n, r_{j_1}^n)$ to $\partial D(p_{j_2}^n, r_{j_2}^n)$ in case $\mathrm{dist}(p_{j_1}^n, p_{j_2}^n) - (r_{j_1}^n + r_{j_2}^n) \to 0$ and similarly in case $i(p_j^n) - r_j^n \to 0$ to apply a similar argument as before. $\qquad \square$

2. Minimal Surfaces of Higher Genus with Free Boundaries

We make the following assumptions about the geometric setting:

(i) X is a three-dimensional manifold of bounded geometry, i.e. the sectional curvature is bounded and the injectivity radius is bounded from below by a positive constant.

(ii) ∂A has nonnegative mean curvature in the sense of Meeks–Yau [12], namely it consists of a finite number of C^2-surfaces $H_1, \ldots, H_m$ with:

   (a) H_i has nonnegative mean curvature with respect to the interior normal;*

   (b) H_i is a compact subset of a smooth surface $\bar{H}_i$ in X with

$$\bar{H}_i \cap A = H_i,$$

$$\partial \bar{H}_i \subset X \backslash A, \qquad i = 1, \ldots, m$$

(A is a closed subset of X).

(iii) $K \subset X$ is a closed subset. ∂K has bounded C^2-norm in the sense of [8] (K may be empty).

(iv) $\Gamma_j, j = 1, \ldots, l$ is a collection of Jordan curves on ∂A, $\Gamma_j \cap K = \varnothing$ for $j = 1, \ldots, l$.

(v) If v_A and v_K denotes the resp. unit normal vectors

$$v_A \cdot v_K \geqslant 0 \quad \text{in } \partial A \cap \partial K. \tag{2.1}$$

At points where two or more pieces of ∂A come together, (2.1) is required to hold for the normal vectors of all these pieces.

We put

$$\mathcal{M}(g) := \{M: M \subset A \cap X \backslash K, M \text{ is a compact oriented embedded surface}$$
$$\text{of genus } g \text{ with}$$
$$\partial M = \bigcup_{j=1}^{l} \Gamma_j \cup \bigcup_{i=1}^{q} \gamma_i, \text{ where } \gamma_i \subset \partial K \text{ are closed Jordan curves}$$
$$(q \geqslant 0; q = 0 \text{ corresponds to the case } \bigcup \gamma_i = \varnothing)\}.$$

$\alpha_g := \{\inf |M|: M \in \mathcal{M}(g)\}$

   (here, $|M|$ denotes the area of M).

If $\mathcal{M}(g) = \varnothing$, we put $\alpha_g = \infty$; likewise $\alpha_{-1} = \infty$.

$$\alpha'_g = \{\inf |M|: M = \bigcup_{\mu=1}^{d} M_\mu, M_\mu \in \mathcal{M}(g_\mu), \bigcup_{j=1}^{l} \Gamma_j \subset \bigcup_{\mu=1}^{d} \partial M_\mu \subset \bigcup_{j=1}^{l} \Gamma_j \cup \bigcup_{i=1}^{q} \gamma_i,$$
$$\text{where } \gamma_i \subset \partial K \text{ are closed Jordan curves, and either } (d \geqslant 1 \text{ and}$$
$$\sum_{\mu=1}^{d} g_\mu < g) \text{ or } (d \geqslant 2 \text{ and } \sum_{\mu=1}^{d} g_\mu \leqslant g)\}.$$

Note

$$\alpha'_g \leqslant \alpha_{g-1}.$$

Finally,

$\mathcal{M}(g, h)$ is the set of those surfaces appearing in the definition of $\mathcal{M}(g)$, where q (the number of free boundary curves) equals precisely h.

$\alpha_{g,h} := \{\inf |M|: M \in \mathcal{M}(g, h)\}$.

---

*Note that we have changed the sign of the mean curvature vector compared to [12] in order to make it agree with the standard sign convention in differential geometry.

$\alpha'_{g,h}$ then is defined similarly to α'_g, where we now take unions of surfaces that either lower the genus, increase the number of components, or decrease the number of free boundary curves.

Theorem 2. *Suppose*

$$\alpha_g < \alpha'_g \text{ (in particular, } \mathcal{M}(g) \neq \varnothing). \tag{2.2}$$

Then there exists an embedded minimal surface M in $X\backslash K$ of genus g, having Γ_j, $j = 1, \ldots, l$, as fixed boundary curves and possibly free boundary curves γ_i, $i = 1, \ldots, p$, on ∂K along which M meets ∂K orthogonally. M is of class $C^{2,\alpha}$ in the interior and is regular at the boundary as Γ_j and ∂K permit. In particular, M is at least of class $C^{1,\alpha}$ near ∂K.

PROOF. The case $g = 0$ was treated in [7; Theorem 6.1]. The arguments of [7] still pertain to the present case. Some additional arguments are required, however, to treat the case of higher genus. For this reason, we sketch those arguments that can be taken from [7] and add those ones necessary for $g > 0$. Without loss of generality we shall actually assume $g \geqslant 2$, since the case $g = 1$ is similar, but easier.

We first assume that $\partial A \in C^2$, ∂A has positive mean curvature with respect to the interior normal, and $\Gamma_j \in C^2$, $j = 1, \ldots, l$.

We take an area minimizing sequence $(\Sigma_n)_{n \in \mathbb{N}}$ in $\mathcal{M}(g)$. We need the following lemma (cf. [1; Lemma 5] and [7; Lemma 6]; the proof of those results can easily be adapted to our present needs).

Lemma 1. *There exists a constant $c = c(X) > 0$, depending only on the geometry of X (curvature bounds and a lower bound for the injectivity radius of X) with the following property:*
    Let $M \in \mathcal{M}(g)$, $|M| < \alpha_{g-1}$, $\rho < c(\alpha_{g-1} - |M|)$, U a convex open set of class C^2 with diameter $\leqslant \rho$. If

$$\partial M \cap U \cap X\backslash K = \varnothing$$

and M intersects ∂U transversally, then for each component σ of $M \cap \partial U$ there exists an embedded disk with holes $N \subset M$ with

$$\partial N \subset \sigma \cup \partial K.$$

We can then proceed as in [7; proofs of Theorems 6.1 and 6.3] and obtain an embedded minimal surface Σ as varifold limit of (Σ_n). $\Gamma_j \subset \partial \Sigma$, and Σ has multiplicity $\frac{1}{2}$ at Γ_j. Also, $|\Sigma| \leqslant \alpha_g$. We discard those components of Σ that have no boundary curve Γ_j. Σ then has multiplicity 1 at interior points and hence is a union of orientable surfaces (cf. [10; Remark (3.27)]), the sum of whose genera does not exceed g. It then follows from our assumption (2.2) that Σ has precisely one component, and this component has genus precisely g. Σ is bounded, as otherwise the area of Σ would be infinite as a consequence of the monotonicity formula and the assumption that X has bounded geometry. Moreover, the number of boundary curves of Σ on ∂K is finite (possibly zero). This is an application of the regularity theorem of [3] for varifolds with free boundaries, cf. [7; proof of Theorem 6.1].

The general case will now be treated by approximation arguments, using constructions of [11] and [12]. First, we pass from $\partial A \in C^2$ with positive mean curvature to ∂A satisfying only (ii). Since Σ is bounded and $\Gamma_j \subset \Sigma$, we can assume without loss of generality that A is compact. Hence, by [12; §1], A can be approximated by a sequence (A_k) of compact smooth manifolds with smooth boundary ∂A_k of positive mean curvature so that the metrics, their derivatives, and the boundaries converge to the resp. objects of A uniformly. Also, each Γ_j is approximated by smooth curves $\Gamma_{j,k} \subset \partial A_k$.

We minimize the area in A_k among surfaces of genus g with fixed boundary curves $\Gamma_{j,k}$ and possibly free boundaries on ∂K. As shown above, we obtain an embedded minimal surface Σ_k of genus g. We let q_k denote the number of free boundary curves of Σ_k. By Theorem 1, there exists an (injective) conformal map $f_k : S_k \to X$, with $f(S_k) = \Sigma_k$, where each S_k can be considered as a hyperbolic surface $S_{o,k}$ with q_k geodesic disks removed. We want to verify that the surfaces $S_{o,k}$ represent a compact subset of M_g, the moduli space of surfaces of genus g. Using Mumford's compactness theorem ([13]), we have to ensure that the lengths of closed geodesics on $S_{o,k}$ stay bounded from below by a positive constant (independent of k). If, however, this were not the case, i.e. the lengths of some closed geodesics tend to zero as $k \to \infty$, we could use the arguments of [6; §2] (based on the collar lemma) to find homotopically nontrivial curves δ_k in S_k for which the image curves $f_k(\sigma_k)$ become arbitrarily short as $k \to \infty$ (of course, the argument is not affected by the possibility that a collar might contain one of the disks removed from $S_{o,k}$ to obtain S_k).

On the other hand, since the metrics converge also, the area of Σ_k with respect to the metric of A approaches α_g. Hence, without loss of generality, for all k,

$$|\Sigma_k| \leqslant \alpha_g - \tfrac{1}{2}(\alpha_g - \alpha_{g-1}) < \alpha_{g-1} \quad \text{by (2.2).}$$

Hence the existence of arbitrarily short closed geodesics on $S_{o,k}$ is not compatible with Lemma 1, and hence (by [13]), $S_{o,k}$ converges to a hyperbolic surface S_o of genus g. Without loss of generality, $S_{o,k} = S_o$ for all k. Moreover, we can also assume that the radii and centers of the q_k disks removed from $S_{o,k}$ converge. Some of the disks may degenerate to points or become tangent to each other in the limit, of course. No degeneration, however, can occur of those disks whose boundary corresponds to one of the fixed boundary curves Γ_j, cf. [6] again, noting $K \cap \Gamma_j = \varnothing$ for $j = 1, \ldots, l$. We also let S denote the limit of the S_k's.

As in [7; §6], we then derive uniform $C^{1,\alpha}$-estimates for $f_k : S_k \to A_k$, implying that (after selection of a subsequence) we get a $C^{1,\alpha}$ limit map

$$f : S \to A.$$

f is weakly harmonic, hence of class $C^{2,\alpha}$ in the interior of S, moreover conformal, hence represents a parametric minimal surface. If some of the disks removed from $S_{o,k}$ shrink to points in the limit, they yield isolated, hence, removable singularities of f. Moreover, $\Sigma := f(S)$, on the other hand, also is the varifold limit of Σ_k, hence embedded (cf. [1] and [6; §6]), as (Σ_k) can also be considered as an area minimizing sequence in $\mathcal{M}(g)$. The regularity of the manifold Σ at the free boundary excludes that some of the disks removed from $S_{o,k}$ become tangent to each other in the limit without shrinking to a point.

Also, Σ has multiplicity $1/2$ at Γ_j again, and hence $f: S \to \Sigma$ cannot be a multiple covering (cf. [10; §3]). Altogether, $\Sigma = f(S)$ is an embedded oriented minimal surface of genus g with fixed boundary curves Γ_j ($j = 1,\ldots,l$) and possibly free boundary curves on ∂K. f, hence Σ, is regular at the fixed boundary by [4], regular at the free boundary by [8], and regular in the interior by [9].

The other approximation arguments can be taken over from [6] (based on [12]). $\square$

In a similar way, one can also show (cf. Theorem 6.2 of [7]):

Theorem 3. *Assume in addition that K has nonpositive mean curvature with respect to the interior normal. Let*

$$\alpha_{g,h} < \alpha'_{g,h} \quad \text{for some } g \geq 0, h \geq 0$$

(in particular, $\mathcal{M}(g,h) \neq \varnothing$). Then there exists an embedded minimal surface of genus g with precisely h free boundary curves on ∂K, satisfying the conclusion of Theorem 2.

Acknowledgment. This research was supported in part by NSF Grant No. 8120790.

References

[1] F. Almgren Jr. and L. Simon, *Existence of embedded solutions of Plateau's problem*, Ann. Sc. N. Pisa (iv), **6** (1979), 447–495.

[2] M. Freedman, J. Hass, and P. Scott, *Least area incompressible surfaces in 3-manifolds*, Inv. Math., **71** (1983), 609–642.

[3] M. Grüter and J. Jost, *Allard type regularity results for manifolds with free boundaries*, Ann. Sci. Norm. Sup. Pisa. In press (1986).

[4] E. Heinz and S. Hildebrandt, *Remarks on minimal surfaces in Riemannian manifolds*, CPAM **23** (1970), 371–377.

[5] J. Jost, *Harmonic maps between surfaces*, Lecture Notes in Mathematics, Springer-Verlag, New York, **1062** (1984).

[6] J. Jost, *Conformal mappings and the Plateau–Douglas problem*, J. Reine Angew. Math., **359** (1985), 37–54.

[7] J. Jost, *Existence results for embedded minimal surfaces of controlled topological type* I, Ann. Sci. Norm. Sup. Pisa. In press (1986).

[8] J. Jost, *On the regularity of minimal surfaces with free boundaries in Riemannian manifolds*, Man. Math. (to appear).

[9] J. Jost and H. Karcher, *Geometrische Methoden zur Gewinnung von a-priori— Schranken für harmonische Abbildungen*, Man. Math., **40** (1982), 27–77.

[10] W. Meeks III, L. Simon, and S.T. Yau, *Embedded minimal surfaces, exotic spheres, and manifolds with positive Ricci curvature*, Ann. Math., **116** (1982), 621–659.

[11] W. Meeks III and S.T. Yau, *The classical Plateau problem and the topology of three-dimensional manifolds*, Top. **21** (1982), 409–442.

[12] W. Meeks III and S.T. Yau, *The existence of embedded minimal surfaces and the problem of uniqueness*, Math. Z., **179** (1982), 151–168.

[13] D. Mumford, *A remark on Mahler's compactness theorem*, Proc. AMS, **28** (1971), 289–294.

[14] R. Schoen and S.T. Yau, *Existence of incompressible minimal surfaces and the topology of three dimensional manifolds with nonnegative scalar curvature*, Ann. Math., **110** (1979), 127–142.

[15] K. Schüffler, *Indextheorie für Minimalflächen vom Geschlecht 1* (to appear).

[16] K. Strebel, *Ein Klassifizierungsproblem für Riemannsche Flächen vom Geschlecht 1* (to appear).

Free Boundaries in Geometric Measure Theory and Applications

Michael Grüter

1

The most famous problem in the theory of minimal surfaces is the so called Plateau problem, where one is looking for a minimal surface spanning a given boundary. This is a problem with a fixed boundary and was essentially solved around 1930 by Douglas and Radó.

If we prescribe (part of) the boundary of the minimal surface to lie in a given submanifold we have an example of a (partially) free boundary problem.

In all these cases it is rather easy to prove existence of a solution if one suitably enlarges the class of competing surfaces. Within the framework of geometric measure theory the appropriate generalization of an oriented submanifold is the notion of integer multiplicity rectifiable current. Let

$$\mathscr{D}^n(U) = \{\omega : \omega \text{ a smooth } n\text{-form, spt } \omega \subset U, \text{ spt } \omega \text{ compact}\}$$

for any open set $U \subset \mathbb{R}^{n+k}$, and denote by

$$\mathscr{D}_n(U) = [\mathscr{D}^n(U)]^*$$

the dual space of *n-dimensional currents* in U. We call $T \in \mathscr{D}_n(U)$ an *integer multiplicity rectifiable current* if

$$T(\omega) = \int_M \langle \omega(x), \zeta(x) \rangle \theta(x) \, d\mathscr{H}^n(x), \qquad \omega \in \mathscr{D}^n(U).$$

Here $M \subset U$ is $\mathscr{H}^n$-measurable, countably n-rectifiable, $\theta \colon M \to \mathbb{N}_0$ is locally $\mathscr{H}^n$-integrable, $\zeta(x) = \tau_1 \wedge \cdots \wedge \tau_n$ for $\mathscr{H}^n$-a.e. $x \in M$, $\{\tau_1, \ldots, \tau_n\}$ an orthonormal basis for $T_x M$ (which exists $\mathscr{H}^n$-a.e.), and $\mathscr{H}^n$ denotes as usual n-dimensional Hausdorff measure.

We write

$$T = \underline{\tau}(M, \theta, \zeta).$$

EXAMPLE. If $M \subset U$ is an oriented n-dimensional C^1-submanifold of locally finite $\mathcal{H}^n$-measure we define $[\![M]\!] \in \mathcal{D}_n(U)$ by

$$[\![M]\!](\omega) = \int_M \omega.$$

For $T \in \mathcal{D}_n(U)$ the boundary $\partial T \in \mathcal{D}_{n-1}(U)$ is defined by

$$\partial T(\omega) = T(d\omega).$$

By Stokes' theorem we have $\partial[\![M]\!] = [\![\partial M]\!]$.

For integer multiplicity currents we have the very strong compactness theorem by Federer and Fleming which may be regarded as a first step towards regularity. Let us recall the known regularity results.

If the codimension k is greater than one we have Almgren's recent result for interior regularity. Off a closed set of codimension at least two, an area minimizing integer multiplicity current is smooth. Since any holomorphic variety is absolute area minimizing, this result is the best possible. Almost nothing is known about boundary regularity (fixed or free boundary).

In the special case of codimension one we have much more information. In a series of papers by De Giorgi, Fleming, Almgren, Simons, and Federer it was shown that the interior singular set has at least codimension 7 (1960–1970).

The 7-dimensional cone

$$C = \{x \in \mathbb{R}^8 : x_1^2 + \cdots + x_4^2 = x_5^2 + \cdots + x_8^2\}$$

provides an example of an area minimizing hypersurface with an isolated singularity, as was first shown by Bombieri, De Giorgi, and Giusti (1969). In 1979 Hardt and Simon proved the regularity at a fixed smooth boundary in all dimensions.

2

Concerning the regularity at the free boundary we have the following.

Theorem 1. *Suppose $S \subset \mathbb{R}^{n+1}$ is a C^2-hypersurface, $U \subset \mathbb{R}^{n+1}$ is open and $\partial S \cap U = \varnothing$.*

Let $T \in \mathcal{D}_n(U)$ be an integer multiplicity current with $\mathrm{spt}\, \partial T \subset S$ and

$$\underline{M}_W(T) \leqslant \underline{M}_W(T + X)$$

for any $W \subset\subset U$ and any integer multiplicity $X \in \mathcal{D}_n(U)$ such that $\mathrm{spt}\, X \subset W$ and $\mathrm{spt}\, \partial X \subset S$.

Then, we have

$$\text{sing } T = \varnothing, \qquad n \leqslant 6$$

$$\text{sing } T \text{ is discrete,} \qquad n = 7$$

$$\dim(\text{sing } T) \leqslant n - 7, \qquad n > 7.$$

If $x \in S \cap \text{reg } T$, then S and $\text{spt } T$ intersect orthogonally in a neighborhood of x. For $n = 7$ the cone $C^+ = C \cap \{x_1 \geqslant 0\}$ and the hyperplane $S = \{x_1 = 0\}$ yield an example with an isolated singularity.

This result can be applied to different situations. For example, consider an oriented $(n - 1)$-dimensional submanifold B (possibly with boundary), the fixed boundary, and a compact C^2-hypersurface S, the supporting surface, with $\partial B \subset S$ and such that B and S intersect transversally. If $\mathscr{C}$ is defined by

$$\mathscr{C}(S, B) = \{T \in \mathscr{D}_n(\mathbb{R}^{n+k}): T \text{ integer multiplicity, spt } T \text{ compact,}$$

$$\text{spt}(\partial T - \llbracket B \rrbracket) \subset S\},$$

one can prove the existence of a minimizer to which Theorem 1 may be applied ($k = 1$) near points $x \in \text{spt } T \cap (S \sim \bar{B})$. It may happen that the interior of the solution surface intersects S or that near the free boundary the surface lies on both sides of S.

We can also prove some results in arbitrary codimension. Consider the problem just described, let $\rho(x) = \text{dist}(x, S)$, denote by $\xi(x)$ the nearest point in S (if defined), and let κ be a curvature bound for S.

Theorem 2 (codim $\geqslant 1$). *For $0 < h < 1/2\kappa$ we have*

$$\underline{M}(\partial T - \llbracket B \rrbracket) \leqslant \underline{M}(T \, \llcorner \, \{\rho < h\})(1/h + \kappa) + Ch,$$

with $C = C(S, B)$. Thus ∂T is also an integer multiplicity rectifiable current.

Furthermore, we show the existence of oriented tangent cones to T at any $x \in S \cap (\text{spt } T \sim \bar{B})$ which are minimizing with respect to the hyperplane $T_x S$.

Let me now briefly indicate how Theorem 1 is proved.

PROOF OF THEOREM 1. First we restrict T to lie on one side of S (additional boundary may be introduced). Reflection across S gives a new current T' which is not minimizing anymore. Since we are in the codimension one case we may apply the decomposition theorem and henceforth assume that the density of T' is almost everywhere equal to one. We want to use Simon's abstract version of the dimension reducing argument. To do that we apply regularity theorems for varifolds by Allard respectively Jost–Grüter for the case of interior regularity respectively the case of the free boundary. See below.   □

Since the proof of Theorem 2 is not too hard, I am going to present it in some detail.

PROOF OF THEOREM 2. The estimate will follow from a monotonicity result. We show that (for some C as in the statement of the theorem)

$$m(h)/h + \kappa m(h) + Ch$$

is increasing. Here

$$m(h) = \underline{M}(T \llcorner \{x: \rho(x) < h\})$$

is the mass of T in a tubular neighborhood of S.

For $h \geqslant 0$ we consider the slice

$$\langle T, \rho, h_+ \rangle = (\partial T) \llcorner \{\rho > h\} - \partial(T \llcorner \{\rho > h\})$$

which is again integer multiplicity for a.e. h. Since $\mathscr{H}^{n-1}(S \cap B) = 0$ and $\mu_T(S) = 0$, we get

$$\langle T, \rho, 0_+ \rangle = [\![B]\!] - \partial T,$$

and in view of Lip $\rho = 1$ we have

$$\underline{M}(\langle T, \rho, 0_+ \rangle) \leqslant \liminf_{h \downarrow 0} m(h)/h$$

as well as

$$\underline{M}(\langle T, \rho, h_+ \rangle) \leqslant m'(h) \tag{1}$$

for a.e. $h \geqslant 0$.

We construct a comparison current as follows by using the homotopy

$$f(t, x) = x + t(\xi(x) - x);$$

i.e. we connect each point x with the nearest point $\xi(x) \in S$. We define three currents

$$T_1(h) = f_\#([\![(0,1)]\!] \times \langle T, \rho, h_+ \rangle)$$
$$T_2(h) = f_\#([\![(0,1)]\!] \times [\![B \cap \{\rho < h\}]\!])$$
$$T_3(h) = T \llcorner \{\rho > h\}$$

and finally

$$T(h) = -T_1(h) - T_2(h) + T_3(h).$$

Since $\partial B \subset S$ and $\mathcal{H}^{n-1}(B \cap \{\rho = h\}) = 0$ for a.e. h, one checks that

$$T(h) \in \mathscr{C}.$$

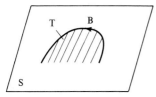   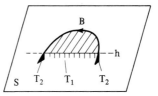

The minimality of T gives

$$m(h) \leqslant \underline{M}(T_1(h)) + \underline{M}(T_2(h)). \tag{2}$$

It is not too hard to see that

$$\underline{M}(T_2(h)) \leqslant Ch^2, \tag{3}$$

because S and B intersect transversely.

To estimate $\underline{M}(T_1(h)) = \sup_{|\omega| \leqslant 1} T_1(h)(\omega)$ we write $R = \langle T, \rho, h_+ \rangle$ and use the formula

$T_1(h)(\omega)$

$$= f_\#([\![(0,1)]\!] \times R)(\omega)$$

$$= \int_0^1 \int_{\mathbb{R}^{n+k}} \langle \omega(f(t,x)), (\xi(x) - x) \wedge (tD\xi(x) + (1-t)\mathbb{1})_\# \vec{R}(x) \rangle \, d\mu_R(x) \, dt,$$

where $\vec{R}(x) = \zeta_R(x) \in \Lambda_{n-1}(\mathbb{R}^{n+k}) \quad \mu_R\text{-a.e}$

A simple calculation yields the estimate ($\kappa h \leqslant 1/2$)

$$\underline{M}(T_1(h)) \leqslant (h + \kappa h^2)\underline{M}(\langle T, \rho, h_+ \rangle). \tag{4}$$

Collecting the estimates (1)–(4) we get

$$m(h) \leqslant m'(h)(h + \kappa h^2) + Ch^2.$$

This is equivalent to

$$[m(h)/h + \kappa m(h) + Ch]' \geqslant 0,$$

and since $m(h)$ is increasing, the desired monotonicity follows.   □

Next, I want to discuss the regularity theorem for varifolds with a free boundary by Jost–Grüter which was mentioned before.

First we have to define the notion of a rectifiable n-varifold. Suppose $M \subset \mathbb{R}^{n+k}$ is $\mathcal{H}^n$-measurable and countably n-rectifiable, $\theta \geqslant 0$ a locally $\mathcal{H}^n$-integrable function on M. The *rectifiable n-varifold* $V = \underline{v}(M, \theta)$ is the equivalence class of all pairs $(\tilde{M}, \tilde{\theta})$ with $\mathcal{H}^n((M \sim \tilde{M}) \cup (\tilde{M} \sim M)) = 0$ and $\theta = \tilde{\theta}$ $\mathcal{H}^n$-a.e. on $M \cap \tilde{M}$.

If $T = \underline{\tau}(M, \theta, \zeta)$ is integer multiplicity, $V = \underline{v}(M, \theta)$ is a rectifiable varifold (with integer multiplicity). Interior regularity and regularity near the fixed boundary for stationary varifolds, in fact for varifolds with generalized mean curvature in L^p, $p > n$, was treated by Allard (1972, 1975).

Theorem 3 (Jost–Grüter). *For any n, $k \in \mathbb{N}$ there are $\gamma(n, k) > 0$ and $\varepsilon(n, k) > 0$ such that the following is true. If $\rho \leqslant 1$, $S \subset \mathbb{R}^{n+k}$ is a C^2-hypersurface, κ a bound for the curvature of S,*

$$0 \in S, \quad \partial S \cap B_1(0) = \phi, \qquad \kappa\rho \leqslant \varepsilon^2,$$

and if $V = \underline{v}(M, \theta)$ is a rectifiable n-varifold ($\mu = \mathscr{H}^n \sqcup \theta$) with

$$0 \in \operatorname{spt} \mu, \quad \theta \geqslant 1 \quad \mu\text{-a.e.}$$

$$\operatorname{spt} \mu \subset \overline{B_1'(0)}$$

(∗) $$\mu(B_\rho(0)) \leqslant \omega_n \rho^n (1 + \varepsilon)/2$$

(∗∗) $$\int \operatorname{div}_M X \, d\mu = 0$$

for all $X \in C_c^1(B_1(0), \mathbb{R}^{n+k})$, $X|_S : S \to TS$, then (after performing a suitable isometry) there is a $C^{1,1/2}$-function $u : B_{\gamma\rho}^n(0) \to \mathbb{R}^k$ such that ($B_{\gamma\rho}^n(0) = B_{\gamma\rho}(0) \cap (\mathbb{R}^n \times \{0\})$)

$$u(0) = 0, \quad v_s(x) \subset T_x \text{ graph } u \text{ for } x \in S \cap \text{ graph } u,$$

$$\operatorname{spt} \mu \cap B_{\gamma\rho}(0) = \text{graph } u \cap B_{\gamma\rho}(0) \cap \overline{B_1'(0)},$$

and

$$\rho^{-1} \sup|u| + \sup|Du| + \rho^{1/2} \sup \frac{|Du(x) - Du(y)|}{|x - y|^{1/2}} \leqslant C\varepsilon$$

with $C = C(n, k)$.

Remarks. The conditions on S are just normalizations. The only condition one has to check before the theorem can be applied to minimal surfaces is (∗). Condition (∗∗) means that $\underline{H} \equiv 0$ (the generalized mean curvature) and that V and S intersect orthogonally. Essentially, the same result is true for $|\underline{H}| \in L^p(\mu)$, $p > n$. The isometry is chosen so that $v_s(0) \subset \mathbb{R}^n = T_0 M$.

PROOF OF THEOREM 3. The main point of the proof is a monotonicity result near the free boundary. One can show that

$$e^{c\kappa r} \omega_n^{-1} r^{-n} [\mu(B_r(x)) + \mu(\sigma(B_r(x)))]$$

is increasing. Here

$$\sigma(x) = 2\xi(x) - x$$

is the reflection across S and ξ, as before, the nearest point projection. From the monotonicity we get the existence of the density

$$\theta^n(\mu, x) = \lim_{r \downarrow 0} \mu(B_r(x))/\omega_n r^n$$

at every $x \in \text{spt } \mu$, and the upper-semicontinuity of

$$\tilde{\theta}^n(\mu, x) = \begin{cases} \theta^n(\mu, x), & x \notin S \\ 2\theta^n(\mu, x), & x \in S. \end{cases}$$

The rest of the proof proceeds, with minor modifications, along the lines of Allard's proof of interior regularity. □

3

Finally, let me just mention a few applications of the above regularity theorems.

 (i) Existence of embedded minimal surfaces with prescribed homology class of the boundary for $n \leqslant 6$ (easy application of Theorem 1).
(ii) Existence of embedded minimal disks in convex bodies (Jost–Grüter). Besides Theorem 3, we use the minimaxing procedure of Pitts, the method of Almgren and Simon to minimize among embedded surfaces, extensions of curvature estimates by Schoen and Simon to prove compactness, and ideas from a paper by Simon and Smith to control the topological type.
(iii) Existence of embedded minimal surfaces of controlled topological type. (Jost, two papers, generalization and extension of the result in (ii)).

References

M. Grüter, *Regularität von minimierenden Strömen bei einer freien Randbedingung*, Habilitationsschrift, Düsseldorf (1985).

M. Grüter, *Regularity Results For Minimizing Currents With A Free Boundary*, Preprint (1985).

M. Grüter, *Optimal Regularity For Codimension One Minimal Surfaces With A Free Boundary*, Preprint (1985).

M. Grüter and J. Jost, *Allard Type Regularity Results For Varifolds With Free Boundaries*, Ann. d. Sc. Norm. Sup. di Pisa (to appear).

M. Grüter and J. Jost, *On Embedded Minimal Disks in Convex Bodies*, Analyse Non Linéaire (to appear).

J. Jost, *Existence Results For Embedded Minimal Surfaces Of Controlled Topological Type I, II*, Preprints 691, 726, SFB 72, Bonn (1984, 1985).

A Mathematical Description of Equilibrium Surfaces

Mario Miranda

The central point in many problems of mathematical physics is answering questions about the boundary of a region, using as little information as possible about the region itself.

As an example, one can compute the measure of the boundary of a subset of the euclidean n-dimensional space once one is able to calculate the integrals of smooth functions over the given set. That is possible thanks to the divergence theorem formula

$$\int_{\Omega} \operatorname{div} \phi(x)\, dx = \int_{\partial\Omega} \phi(x) \cdot v\, dH_{n-1} \tag{1}$$

which yields to

$$H_{n-1}(\partial\Omega) = \sup \left\{ \int_{\Omega} \operatorname{div} \phi(x)\, dx \,\big|\, \phi \in [C_0^1(R^n)]^n, |\phi(x)| \leq 1 \ \forall x \right\} \tag{2}$$

in all cases when the boundary $\partial\Omega$ is sufficiently smooth in order to be able to establish the divergence theorem and to compute $H_{n-1}(\partial\Omega)$.

The identity (2) can be used as a definition of the $(n-1)$-measure for $\partial\Omega$ for all Lebesgue measurable sets Ω. That was done in 1954 by Ennio De Giorgi in "Su una teoria generale della misura $(r-1)$-dimensionale in uno spazio ad r dimensioni" (see Vol. 36 of Annali di Matematica pura e applicata, published in 1954). The same author established many interesting properties of Lebesgue measurable sets whose boundaries have a finite $(n-1)$-measure in the sense of (2) (see the article "Nuovi teoremi relativi alle misure $(r-1)$-dimensionali in uno spazio ad r dimensioni", in Vol. 4 of Ricerche di Matematica, published in 1955).

De Giorgi himself was able to state and prove the isoperimetric property of euclidean balls in the following way:

"For any Lebesgue measurable set Ω of R^n, for $n \geq 2$, the $\min\{\text{meas}\,\Omega,$ $\text{meas}(R^n - \Omega)\}$ is less than or equal to

$$c(n)\left[\sup\left\{\int_\Omega \operatorname{div}\phi(x)\,dx \,|\, \phi \in [C_0^1(R^n)]^n, |\phi(x)| \leq 1 \ \forall x\right\}\right]^{n/n-1},$$

where

$$c(n) = [\omega(n)n^n]^{-1:(n-1)},$$

with

$$\omega(n) = \text{meas}(\{x \in R^n | |x| < 1\}).\text{"}$$

(See the article "Sulla proprietà isoperimetrica della ipersfera, nella classe degli insiemi aventi frontiera orientata di misura finita", in Vol. 5 of Memorie dell'Accademia Nazionale dei Lincei, published in 1958).

Another successful application of the boundary measure definition (2), confirming the power of the new approach, was the solution of Plateau Problem for codimension one surfaces in euclidean spaces. Again E. De Giorgi in 1960 was able to prove the following result:

"If E is an n-dimensional Lebesgue measurable set with minimal boundary in the open set A of R^n, then there exists an open subset A_0 of A such that

(i) $\partial E \cap A_0$ is an $(n - 1)$-dimensional analytic manifold,
(ii) $H_{n-1}(A - A_0) = 0$." Later on, Herbert Federer was able to improve (ii) by showing that
(iii) $H_s(A - A_0) = 0 \quad \forall$ real $s > n - 8$.

De Giorgi–Federer's theorem was extended by Umberto Massari to a larger class of minimal boundaries, including cases of great interest in mathematical physics, such as capillary surfaces. Massari's theorem can be stated as follows:

"If A is an open set of R^n and $H \in L_{\text{loc}}^p(A)$ with $p > n$, if E is a Lebesgue measurable set such that

$$H_{n-1}(K \cap \partial E) < \infty, \quad \forall \text{ open } K \subset\subset A, \tag{α}$$

$$H_{n-1}(K \cap \partial E) + \int_{K\cap E} H(x)\,dx \leq H_{n-1}(K \cap \partial F) + \int_{K\cap F} H(x)\,dx, \tag{β}$$

$\forall$ Lebesgue measurable sets F with $(F - E) \cup (E - F) \subset\subset K$, then there exists an open subset A_0 of A such that

$$\partial E \cap A_0 \text{ is a } (n - 1)\text{-dimensional manifold of class } C^1, \tag{γ}$$

$$H_s(A - A_0) = 0 \quad \forall \text{ real } s > n - 8.\text{"} \tag{δ}$$

This result of Massari is proven in the article "Esistenza e regolarità delle ipersuperfici di curvatura media assegnata in R^n" (see Vol. 55 of Archive for Rational Mechanics and Analysis, published in 1974).

Another extension of the De Giorgi regularity method was its application to the study of minimal surfaces in the presence of an obstacle. In 1971 I was able to prove the following result:

"If A and B are open sets of R^n and E is a Lebesgue measurable set with

(iv) $\partial E \cap A \neq \emptyset$,
(v) $\partial E \cap B = \emptyset$;

if $\partial B \cap A$ is of class C^1, and the boundary of E is minimal with respect to modifications contained in $A - B$, then ∂E is an $(n - 1)$-dimensional manifold of class C^1 in a neighborhood of $\partial B \cap A$." (See Vol. 16 of Annali dell'Università of Ferrara, published in 1971.)

I think that there is almost no need of emphasizing how these regularity results are useful in mathematical physics, since they apply to minimal sets which are only required to be measurable. The existence of measurable minimal sets can be easily established, once one remarks that the functional

$$\sup \left\{ \int_\Omega \operatorname{div} \phi(x) \, dx \, | \, \phi \in [C_0^1(R^n)]^n, |\phi(x)| \leqslant 1 \; \forall x \right\}$$

is lower semicontinuous with respect to the L^1_{loc}-convergence for Ω.

Other relevant continuity properties of minimal sets are the following ones:

(a) If $\{E_j\}$ is a sequence of minimal sets, L^1_{loc}-converging to a set E, then E is a minimal set.
(b) If in addition to the hypothesis of (a) one has

$$x_j \in \partial E_j, \quad x_j \to x,$$

then one can prove that $x \in \partial E$.
(c) If in addition to the hypotheses of (a) and (b) one has: x is a regular point for ∂E, then one can prove that x_j are regular points of ∂E_j, for large j, and the normal unit vector to ∂E_j at x_j converges to the normal unit vector to ∂E at x.

All these continuity results can be extended to the case when the minimum property is taken, in the sense of Massari, with respect to an integrable function $H \neq 0$. In this case we should assume the sequence of functions H_j to be converging to some function H.

The most remarkable applications of these methods are the proofs of the existence of the equilibrium configuration for hanging or rotating liquid drops.

The first problem was considered and solved by E.H. Gonzalez, U. Massari, and I. Tamanini in their article, "Existence and regularity for the problem of a pendant liquid drop" published by Pacific Journal of Mathematics in Vol. 80 (2).

The three authors considered a liquid drop filling a region E of R^{n+1}, hanging from the fixed horizontal reference plane

$$\{(x, t) | x \in R^n, t = 0\}.$$

The global energy of the drop is given by

$$\mathscr{F}(E) = \int_{t<0} |D\phi_E| + v \int_{t=0} \phi_E \, dx + k \int_{t<0} t\phi_E(x,t) \, dx \, dt,$$

where $\phi_E = \phi_E(x,t)$ is the characteristic function of the set E and

$$\int_{t<0} |D\phi_E|$$

is the total variation of the gradient of ϕ_E in the half space

$$\{(x,t)|x \in R^n, t < 0\}.$$

One cannot expect an absolute minimum to exist for $\mathscr{F}$ if $k > 0$, so the right problem to look at is the existence of local minima. For that purpose we introduce a ground floor

$$\{(x,t)|x \in R^n, T < t < 0\}$$

as an arbitrary lower bound for the configuration E. Assuming the measure of E to be equal to one, and the parameter v to belong to the interval $[-1, 1]$, one can consider, for all sets

$$E \subset \{(x,t)|x \in R^n, T < t < 0\},$$

with meas $E = 1$, the set E_k minimizing $\mathscr{F}$. As $k \to 0 + t$, the sets E_k converge to the set E_0 minimizing

$$\int_{t<0} |D\phi_E| + v \int_{t=0} \phi_E \, dx$$

under the condition: meas $E = 1$.

Disregarding the condition

$$E \subset \{(x,t)x \in R^n, T < t < 0\},$$

the set E_0 must be a portion of a ball, then it is strictly contained in the strip

$$\{(x,t)|x \in R^n, T/2 < t < 0\},$$

if T is conveniently large.

At this point the right question to ask is whether we are able or not to improve the convergence $E_k \to E_0$. Observe that we know

$$\text{meas } E_k \cap \{(x,t)|x \in R^n, T < t < T/2\} \to 0.$$

If we were able to derive from that

$$E_k \cap \{(x,t)|x \in R^n, T < t < T:2\} = \varnothing$$

for small k's, we could conclude that E_k are unrestrained local minima for the same values of k.

The same kind of question arises in the study of rotating drops as approached by S. Albano and E.H. Gonzalez in their article "Rotating drops", published by Indiana University Mathematical Journal in 1983.

Albano–Gonzalez proved the existence of local minima for the energy

$$\mathcal{G}_\varepsilon(E) = \int |D\phi_E| - \varepsilon \int_E |y|^2 \, dy \, dz,$$

where

$$x = (y, z), \quad y \in R^n, \quad z \in R,$$

and E satisfies the conditions

$$\text{meas } E = 1, \quad \int_E x \, dx = 0.$$

Restricting ourselves to sets E contained in the ball

$$B_R = \{ x \in R^{n+1} \,||x| < R \},$$

where R is large enough for

$$\text{meas } B_{R/2} > 1,$$

we can easily prove the existence of E_ε minimizing $\mathcal{G}_\varepsilon$.

For $\varepsilon \to 0 +$ the sets E_ε converge to the ball E_0 centered at 0, with measure equal to one. Therefore we have

$$\text{meas}(E_\varepsilon - B_{R:2}) \to 0.$$

If we were able to prove that this property implies

$$E_\varepsilon - B_{R:2} = \varnothing$$

for small ε's, we would get E_ε to be a local minimum for $\mathcal{G}_\varepsilon$, for the same values of ε.

Albano and Gonzalez showed in their article how the use of the isoperimetric properties of the balls in euclidean spaces, together with the minimum properties of E_ε, are sufficient to prove

$$E_\varepsilon - B_{R:2} = \varnothing$$

for small values of ε.

A similar type of proof was presented in the article by Gonzalez, Massari, and Tamanini for the case of hanging drops.

Interfaces of Prescribed Mean Curvature

I. Tamanini

Several questions of mathematical and physical interest lead to the consideration of an "energy functional" of the following type:

$$F[V] = \text{(weighted area of } S) + \int_V H \, dv, \qquad (*)$$

where S is the surface bounding the region V of n-space and H is a given
summable function. In the following, we shall be concerned with a problem
of this type, representing in a sense a simplified physical situation, and investigate some basic properties of its solutions. The results we obtain may serve
both as an illustration of the use of certain variational techniques and as an
instance of results that could be obtained, under appropriate conditions, in
more general cases.

In order to explain the physical meaning of the problem to be discussed in
the sequel, we briefly recall a typical example arising in the study of capillary
surfaces.

Let us consider a container C in 3-space, partly filled by a liquid V subject
to both surface forces (acting on the interface separating the liquid V from the
surrounding gas $C - V$, and on the contact surface between V and the walls
∂C of the container) and volume forces (such as gravity, kinetic forces, etc.).

The equilibrium configurations of the liquid will then correspond to critical
points of a functional like $(*)$, in which the first contribution (surface energy)
is the area of $S = \partial V$, weighted by different coefficients corresponding to
different portions of S (free interface, contact region; we may also count the
area of S differently, depending on the orientation, as in the study of crystals),
while the second term represents the volume energy of the system.

Of course, if we are interested in least energy configurations, we have to
minimize $F[V]$ subject to some "natural" constraints, such as

$$V \subset C$$

$$|V| = \text{meas } V = \text{given positive constant.}$$

A lot of work has been done on this problem, from different points of view, and with more or less stringent assumptions. (See the references at the end of the paper for some representative papers on the subject.) Nevertheless, a completely satisfactory answer is known only for a limited number of particular (though interesting) cases.

Motivated by the preceding discussion, we will consider in the following section a *model* problem, in which all additional difficulties (caused, e.g. by the actual shape of the container or by the presence of constraints and boundary conditions) are intentionally avoided.

1. The Model Problem

Given a function $H(x)$, defined and summable on R^n ($n \geq 2$), we are interested in finding global (absolute) minimizers of the functional

$$F_H[V] = |D\varphi_V|(R^n) + \int_V H(x)\,dx, \tag{1}$$

that is, measurable set $\Omega \subset R^n$ s.t. $F_H[\Omega] \leq F_H[V]$ for every measurable $V \subset R^n$.

As is well-known, the quantity $|D\varphi_V|(R^n)$, sometimes called "the perimeter of V", is the total variation of the gradient of the characteristic function of V: it represents, in a generalized sense, the area of the boundary ∂V of V (see, e.g. [7] or [11]).

We want to discuss some main points concerning the existence and regularity of minimizers of (1) and their behavior under changes of H; in the next section, we shall consider an "inverse problem", i.e. the problem of *finding H* s.t. a *given* set Ω minimizes (1).

First of all, assuming that a minimizer Ω of (1) has already been found, that $S = \partial\Omega$ is smooth near one of its points y, and that H is continuous at that point, we obtain from the first variation of (1) that the mean curvature of S at y equals $-H(y)/(n-1)$. Here we assume that the mean curvature of a convex surface (the boundary of a convex set in R^n) is non-negative. This is the reason why the minimizers of (1) are usually called "sets of prescribed mean curvature H".

Now we describe the use of direct methods in proving the existence of minimizers of (1). To this aim, we may slightly relax the hypothesis on H. Specifically, we shall establish the result under the assumption that H be an extended valued measurable function s.t. H^- (the negative part of H) $\in L^1(R^n)$. This is important in applications, since clearly by allowing the value $+\infty$ to H one forces the solution to avoid certain regions, thus creating obstacles and artificial containers. We have then

$$F_H[V] \geqslant - \int_{R^n} H^- \, dx > -\infty \quad \forall V \subset R^n$$

while $F_H[\phi] = 0$. Moreover

$$|D\varphi_V|(R^n) \leqslant c + \int_{R^n} H^- \, dx$$

as soon as V satisfies

$$F_H[V] \leqslant c.$$

From known compactness and semicontinuity results, a subsequence $\{V_j\}$ of a minimizing sequence will locally converge toward a solution Ω:

$$V_j \to \Omega \text{ in } L^1_{\text{loc}}(R^n)$$

$$F_H[\Omega] \leqslant \liminf_j F_H[V_j] = \min_V F_H[V].$$

Obviously

$$|D\varphi_\Omega|(R^n) = p < +\infty \tag{2}$$

so that, by virtue of the isoperimetric inequality

$$\min\{|\Omega|, |-\Omega|\} \leqslant c(n) p^{n/(n-1)}, \tag{3}$$

either Ω or its complement $-\Omega$ has finite measure. Notice that if $H \in L^1(R^n)$, then

$$\Omega \text{ minimizes } F_H \text{ iff } -\Omega \text{ minimizes } F_{-H}.$$

It may be interesting to know conditions ensuring that the solution Ω is actually a *bounded* set, this being reasonably expected in many applications. To this aim, one usually imposes suitable "growth conditions" on H at infinity. Indeed, a simple "sign condition" yields the following alternative: suppose that $H \geqslant 0$ outside a certain ball $B_r \subset R^n$. Then

$$\text{either } \Omega \subset B_r \quad \text{or} \quad |\Omega| = +\infty. \tag{4}$$

For the proof, we observe that $|\Omega| < +\infty$ implies

$$|D\varphi_{\Omega \cap B_r}|(R^n) \leqslant |D\varphi_\Omega|(R^n) \tag{5}$$

with equality iff $\Omega \subset B_r$ (in measure, i.e. iff $|\Omega - B_r| = 0$: in our treatment we will not distinguish between equivalent sets). The validity of (5) has been established in [14], Lemma 8, in case Ω is bounded, and follows by approximation in the general case $|\Omega| < +\infty$; see also [4], Corollary 1.4, and inequality (1.1).

Furthermore,

$$\int_{\Omega \cap B_r} H \, dx \leqslant \int_\Omega H \, dx \tag{6}$$

since $H \geqslant 0$ on $-B_r$. By adding (5) and (6) we get

$$F_H[\Omega \cap B_r] \leqslant F_H[\Omega] = \min F_H$$

and thus necessarily $\Omega \subset B_r$, by the remark following (5). Both alternatives (4) can, in fact, occur with *different* solutions corresponding to the *same* function H, as simple examples show. Obviously, if $H \geqslant \varepsilon > 0$ on $-B_r$, then every minimizer of F_H is contained in B_r (otherwise, $F_H[\Omega] = +\infty$, by (4)).

The solution Ω is not unique in general, not even when H is continuous and strictly convex on R^n. We have instead the following useful monotonicity result (see [14], Theorem 5, for the proof of a similar result):

$$\text{if } H_1 < H_2 \quad \text{then } \Omega_2 \subset \Omega_1,$$

where, of course, Ω_i is a minimizer of F_{H_i}, $i = 1, 2$. The same conclusion can be drawn under the assumption $H_1 \leqslant H_2$, *provided* Ω_i ($i = 1$ or 2) is known to be the unique minimizer of F_{H_i}. One can also conceive that a convex function H always gives rise to convex solution sets Ω: to our knowledge, this fact has not yet been proven (nor disproven).

We now discuss the regularity of the solutions. Suppose that Ω is a minimizer of F_H and, for a given ball $B_\rho \subset R^n$, call Ω_ρ a set coinciding with Ω in $-B_\rho$ and having least area boundary in B_ρ. In other words, Ω_ρ is a solution to the following Plateau problem:

$$|D\varphi_V|(\bar{B}_\rho) \to \min, \quad \text{with } V - B_\rho = \Omega - B_\rho.$$

Being $F_H[\Omega] \leqslant F_H[\Omega_\rho]$, we get

$$|D\varphi_\Omega|(B_\rho) - |D\varphi_{\Omega_\rho}|(B_\rho) \leqslant \int_{B_\rho} |H| \, dx \leqslant \text{const.} \, \|H\|_{L^p(B_\rho)} \cdot \rho^{n(1-1/p)} \qquad (7)$$

if $H \in L^p_{loc}(R^n)$, $p > 1$. In particular, when $p > n$, the difference in (7) is estimated by a constant times $\rho^{n-1+2\alpha}$, where $\alpha = (p - n)/2p \in (0, 1/2)$. It follows from the results in [12] that in this case $S = \partial\Omega$ turns out to be a regular hypersurface:

$$S = \partial\Omega \in C^{1,\alpha}$$

at least when the dimension $n \leqslant 7$.

When $n \geqslant 8$, it is well-known that singularities may appear, the Hausdorff dimension of the singular set being in any case $\leqslant n - 8$ (see [12] again).

When $1 < p < n$, singular points (e.g. cusps) may appear even in low dimension, and indeed when $p = 1$ any kind of singular solution is permitted, which is compatible with the requirement of being a set of finite perimeter (recall (2)): this will be shown in the next section. The borderline case $p = n$ still constitutes an open problem: we do not know whether any minimizer of F_H, $H \in L^n$, is regular, nor do we know counterexamples to regularity.

We conclude this section with the following continuity result: assume that $\{H_j\}$ be a sequence of functions of $L^1(R^n)$ s.t.

$$H_j \to H \text{ in } L^1(R^n),$$

and denote by $\{\Omega_j\}$ a corresponding sequence of solutions s.t.

$$\Omega_j \to \Omega \text{ in } L^1(R^n);$$

then Ω is a minimizer of F_H. The proof is quite easy. Much deeper results of the regularity theory (see [12]) are needed in proving that if the convergence $H_j \to H$ takes place in $L^p(R^n)$, $p > n$, then $S_j = \partial\Omega_j$ converges toward $S = \partial\Omega$ in a stronger sense. For example, if y is a regular point of S and $y_j \to y$ with $y_j \in S_j \forall_j$, then it can be shown that S_j is regular near y_j for j big enough, and that the normals to S_j at y_j converge to the normal to S at y.

It is worth noticing that most of the indicated results can be and indeed have been extended to more general variational problems, using essentially the same ideas: the semicontinuity of the energy functional, suitable controls of the type (7), and so on.

2. The Inverse Problem

As a matter of fact, the model problem considered in the preceding section is, in a sense, the most general variational problem concerning free interfaces. The point is that *any* surface $S = \partial\Omega$ of finite area can be viewed as a solution to the model problem for a suitable choice of the function $H \in L^1(R^n)$. In other words, for every set Ω of finite perimeter in R^n, a function H can be found s.t. $H \in L^1(R^n)$ and Ω is a minimizer of F_H.

This fact, which seems to be of considerable importance, has been established by E. Barozzi, E. Gonzalez, and the author (see [3]), and then applied to the study of some questions regarding a certain "penalized" formulation of the obstacle problem for minimal boundaries (see [2]). Its proof might be best illustrated by the following explicit construction.

Consider a container Ω in 3-space, having the shape of an ice-cream cone (figure 1), and call Ω_1 the largest ball contained in Ω and $h_1 > 0$ the (constant) mean curvature of its boundary S_1. We may think of Ω_1 as a soap bubble inside Ω. Blowing up the bubble a little bit, a new configuration $\Omega_2 \supset \Omega_1$ will be reached: call $h_2 > h_1$ the constant mean curvature of the spherical cap $S_2 = \partial\Omega_2 \cap \Omega$.

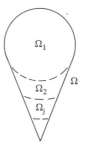

FIGURE 1

Proceeding in this way, an expanding sequence of domains $\{\Omega_j\}$ is produced, which fill completely the container Ω; the corresponding increasing sequence $\{h_j\}$ can then be used to define H on the difference sets $\Omega_j - \Omega_{j-1}$ ($\Omega_0 = \phi$).

In general, assume that $|D\varphi_\Omega|(R^n) < +\infty$ and $|\Omega| < +\infty$ (recall (2) and (3) above). For $\lambda > 0$ and $G \subset \Omega$ define

$$F_\lambda[G] = |D\varphi_G|(R^n) + \lambda|\Omega - G|.$$

Then as λ increases to $+\infty$, the various minimizers G_λ of F_λ will take the place of the expanding soap bubbles of the preceding physical description. Indeed, we can prove that G_λ increases with λ and fills Ω as $\lambda \to +\infty$. Defining for $j \geqslant 1$

$$H^-(x) = \begin{cases} j & \text{on } G_j - G_{j-1} \\ 0 & \text{on } -\Omega \end{cases},$$

where $G_0 = \phi$, yields the negative part of H; in fact, we can show that $H^- \in L^1(R^n)$ and that

$$F_{-H^-}[\Omega] \leqslant F_{-H^-}[V] \; \forall V \subset \Omega. \tag{8}$$

Essentially in the same way we proceed on $-\Omega$ (notice that $|-\Omega| = +\infty$, so that the definition of F_λ has to be conveniently modified! See [3] for details), obtaining $H^+ \in L^1(R^n)$, $H^+ \geqslant 0$, $H^+ = 0$ on Ω, s.t.

$$F_{-H^+}[-\Omega] \leqslant F_{-H^+}[V] \; \forall V \subset -\Omega. \tag{9}$$

Combining (8) and (9), we see easily that $H = H^+ - H^-$ meets our requirements, i.e.

$$H \in L^1(R^n) \text{ and } F_H[\Omega] \leqslant F_H[V] \; \forall V \subset R^n.$$

3. Concluding Remarks

The preceding construction works for general domains. In particular cases, however, the integrand H can be constructed by appropriately extending the normals to $S = \partial\Omega$.

For example, if $\Omega = B_R$ (n-ball of radius R) then $N(x) = x/R$ is a regular extension of the unit normals to ∂B_R, and $H^-(x) = \text{div } N(x) = n/R$ on B_R, $H^- = 0$ on $-B_r$, can be used for the negative part of H.

Arguing in this way, one can show that when Ω is open and bounded with $\partial\Omega$ of class $C^{1,1}$, then H can be chosen to be bounded (and summable) on R^n. Similarly, when $\partial\Omega$ is of class C^2, then a continuous integrand can be found by locally extending the exterior unit normals to $\partial\Omega$.

Some indications in this direction are contained in [2].

References

[1] S. Albano and E. Gonzalez, *Rotating drops*, Indiana Univ. Math. J. **32** (1983), 687–702.

[2] E. Barozzi and I. Tamanini, *Penalty methods for minimal surfaces with obstacles* (to appear).

[3] E. Barozzi, E. Gonzalez, and I. Tamanini, *The mean curvature of a set of finite perimeter* (to appear).

[4] R.C. Bassanezi and I. Tamanini, *Subsolutions to the least area problem and the "minimal hull" of a bounded set in R^n*, Ann. Univ. Ferrara **30** (1984), 27–40.

[5] R. Finn, *Capillarity phenomena*, Uspehi Math. Nauk. **29** (1974), 131–152.

[6] E. Giusti, *The equilibrium configuration of liquid drops*, J. Reine Angew. Math. **321** (1981), 53–63.

[7] E. Giusti, *Minimal surfaces and functions of bounded variations*, Birkhäuser, Boston, 1984.

[8] E. Gonzalez and I. Tamanini (Ed.), *Variational methods for equilibrium problems of fluids* (Proceeding of a Conference held in Trento, 20–25 June 1983), Asterisque **118** (1984).

[9] E. Gonzalez, U. Massari, and I. Tamanini, *Existence and regularity for the problem of a pendent liquid drop*, Pacific J. Math. **88** (1980), 399–420.

[10] U. Massari, *Esistenza e regolarita' delle ipersuperfici di curvatura media assegnata in R^n*, Arch. Rat. Mech. Anal. **55** (1974), 357–382.

[11] U. Massari and M. Miranda, *Minimal surfaces of codimension 1*, North-Holland, Amsterdam, 1984.

[12] I. Tamanini, *Regularity results for almost minimal oriented hypersurfaces in R^n*, Quaderni del Dipartimento di Matematica dell'Universita' di Lecce, N. 1, 1984.

[13] J. Taylor, *Existence and structure of solutions to a class of nonelliptic variational problems*, Symposia Math. **14** (1974), 499–508.

[14] T. Vogel, *Unbounded parametric surfaces of prescribed mean curvature*, Indiana Univ. Math. J. **31** (1982), 281–288.

On the Uniqueness of Capillary Surfaces

Luen-fai Tam

§1

Let $\Omega \subset \mathbb{R}^n$. Consider the equation of prescribed mean curvature

$$\text{div } Tu = H \quad \text{in } \Omega \tag{1}$$

where

$$Tu = \frac{Du}{\sqrt{1 + |Du|^2}} \tag{2}$$

and Du is the gradient of u.

For $H \equiv 0$, $n \leqslant 7$, and $\Omega = \mathbb{R}^n$, Bernstein's theorem says that u must be linear. There is no direct generalization of the theorem to nonzero constant mean curvature H, because in this case there is no solution of (1) defined on the whole space $\mathbb{R}^n$. In fact, Finn [3] proved that if $H = n$, Ω contains the unit ball B_1 of $\mathbb{R}^n$, and (1) has a solution u, then Ω has to be exactly B_1 and u must be a lower hemisphere. Since no boundary condition is imposed, the above result can be viewed as an analogue of Bernstein's theorem. Later in [7], Giusti studied similar problems for general $H = H(x)$ and for general *bounded* domains Ω. He introduced the idea of extremal domain, i.e., a largest domain so that (1) has a solution. He showed that in an extremal domain the solution of (1) is unique up to an additive constant. Furthermore, even though we do not impose any boundary condition, every solution of (1) in an extremal domain in fact satisfies

$$Tu \cdot v = 1, \tag{3}$$

where v is the unit outward normal of Ω. Equation (3) is exactly the type of boundary condition of capillary problem.

Except for the original Bernstein's theorem, little seems to be known about unbounded domains. However, it was conjectured by Finn [3] that the only solution of (1) with $H = 2$ over an infinite strip of width 1 in $\mathbb{R}^2$ is a cylinder. Because of Giusti's result on the boundary behavior of solution of (1) in an extremal domain, we are interested in the following problem related to Finn's conjecture:

$$\begin{cases} \text{div } Tu = H & \text{in } \Omega \\ Tu \cdot v = \cos \gamma & \text{on } \partial\Omega, \end{cases} \tag{4}$$

where H and γ are constants, $\pi/2 > \gamma \geqslant 0$. Ω is the infinite strip $\mathbb{R} \times (-\frac{1}{2}, \frac{1}{2})$ in $\mathbb{R}^2$. We want to study the uniqueness of solutions of (4). Note that it was proved in [13] that if (4) has a solution then $H = 2\cos\gamma$.

Following [3], it is easy to see that the cylinder

$$\Phi(x^1, x^2) = -\sqrt{\left(\frac{1}{2\cos\gamma}\right)^2 - (x^2)^2} \tag{5}$$

is a solution of (4). So is every rigid rotation $\Phi_\beta(x^1, x^2)$ of $\Phi(x^1, x^2)$, $|\beta| < 1$, defined by

$$\Phi_\beta(x^1, x^2) = -\frac{1}{\sqrt{1 - \beta^2}}\left(\sqrt{\left(\frac{1}{2\cos\gamma}\right)^2 - (x^2)^2}\right) + \frac{\beta}{\sqrt{1 - \beta^2}} x^1. \tag{6}$$

Therefore, solution of (4) is not unique up to an additive constant. However, it was proved by the author [13], that any solution of (4) is of the form $\Phi_\beta + $ constant for some $|\beta| < 1$ if $\frac{\pi}{2} > \gamma > 0$. By oral communication we have known that the above statement is still true for $\gamma = 0$ and is proved by Wong [15]. His method does not work for $\gamma > 0$.

In this paper, we want to modify the method in [13] to give a unified proof for all cases $\frac{\pi}{2} > \gamma \geqslant 0$.

§2

In the rest of the paper, Ω always denotes the infinite strip $\mathbb{R} \times (-\frac{1}{2}, \frac{1}{2})$ in $\mathbb{R}^2$.

Main Theorem. *Every solution of (4) is of the form $\Phi_{\beta*} + $ constant for some $|\beta*| < 1$.*

As we noted before, we may assume that $H = 2\cos\gamma$.

To prove the Main Theorem, we will use the idea of generalized solution by M. Miranda [11] (see also [8]). We will also use a well known comparison theorem by Concus and Finn [2].

Consider the following functional

$$F(W) \equiv \int_{\Omega \times \mathbb{R}} |D\chi_w| + \int_{\Omega \times \mathbb{R}} H\chi_w \, dx \, dt - \cos\gamma \int_{\partial\Omega \times \mathbb{R}} \chi_w \, dH_2,$$

where $W \subset \Omega \times \mathbb{R}$ is a Caccioppoli set. χ_w is the characteristic function of W and H_2 is the two-dimensional Hausdorff measure.

Definition 1. A Caccioppoli set U is said to minimize $F(W)$ if and only if for any compact set $K \subset \mathbb{R}^3$ and any Caccioppoli set V of $\Omega \times \mathbb{R}$ such that $\mathrm{spt}(\chi_U - \chi_V) \subset K$, we have $F_K(U) \leqslant F_K(V)$ where

$$F_K(W) \equiv \int_{(\Omega \times \mathbb{R}) \cap K} |D\chi_w| + \int_{(\Omega \times \mathbb{R}) \cap K} H\chi_w \, dx \, dt - \cos \gamma \int_{(\partial\Omega \times \mathbb{R}) \cap K} \chi_w \, dH_2.$$

Definition 2. A measurable function $u \colon \Omega \to [-\infty, \infty]$ is said to be a *generalized solution* of (4) if its subgraph $U = \{(x, s) \in \Omega \times \mathbb{R} | s < u(x)\}$ minimizes $F(W)$.

Definition 3. A sequence of measurable function $u_j \colon \Omega \to [-\infty, \infty]$ is said to converge *locally* to a function u if the characteristic functions of the subgraphs of u_j converge almost everywhere to the characteristic function of the subgraph of u.

One important property of generalized solutions is the following compactness theorem (see [11], see also [8]).

Theorem 1. *Any sequence of generalized solutions of* (4) *contains a subsequence which converges locally to a generalized solution of* (4).

For $(x_0, t_0) \in \mathbb{R}^2 \times \mathbb{R}$ and $r > 0$, define

$$C_r(x_0, t_0) = \{(x, t) \in \mathbb{R}^2 \times \mathbb{R} | |x - x_0| < r \text{ and } |t - t_0| < r\}.$$

For any set U in $\mathbb{R}^3$, define

$$U_r(x_0, t_0) = C_r(x_0, t_0) \cap U \quad \text{and}$$

$$U_r'(x_0, t_0) = C_r(x_0, t_0) - U.$$

From the proof of Theorem 3.2 in [8] and the fact that $\cos \gamma > 0$, we have the following estimates.

Theorem 2. *There exist constants* $r_1 > 0$ *and* $C_1 > 0$ *such that for any generalized solution* u *of* (4) *with subgraph* U *and for any* $(x, s) \in \bar{\Omega} \times \mathbb{R}$, *the following is true:*

if $|U_r'(x, s)| > 0$ *for all* $r > 0$ *then* $|U_r'(x, s)| \geqslant C_1 r^3$ *for all* $r \leqslant r_1$,

where $|W|$ *denotes the Lebesgue measure of the set* W.

Moreover, if $x \in \mathbb{R} \times \{0\}$ then the following is also true:

if $|U_r(x, s)| > 0$ for all $r > 0$, then $|U_r(x, s)| \geqslant C_1 r^3$ for all $r \leqslant r_1$.   (7)

§3

For $t > 0$, let $\Omega_t = \{(x^1, x^2) \in \Omega \mid |x^1| < t\}$. Let $BV_{loc}(\Omega)$ be the set of functions w so that $w|_{\Omega_t} \in BV(\Omega_t)$ for all $t > 0$, where $BV(\Omega_t)$ is the space of functions of bounded variation in Ω_t. Our definition of $BV_{loc}(\Omega)$ is different from the one commonly used.

Lemma 1. *If u is a solution of (4) then u is bounded in Ω_t for every $t > 0$.*

PROOF. Using the method in §3.2 of [2], one can prove that u is bounded from above in Ω_t. Using the boundary condition $Tu \cdot v = \cos \gamma > 0$, one can show that u is bounded below in $\{-t, t\} \times (-\frac{1}{2}, \frac{1}{2})$. Then we can use Φ + constant as comparison surface and use the comparison theorem Theorem 6 by Concus–Finn in [2] to conclude that u is bounded from below in Ω_t, where Φ is the cylinder defined in (5). ☐

By Lemma 1, it is easy to see that $u \in BV_{loc}(\Omega)$. By a theorem of M. Miranda [11] (see also [8]), we can prove that every solution u of (4) is also a generalized solution of (4). Using Theorem 1 and 2 as in [13], one can prove the following.

Lemma 2. *Let u be a solution of (4) and let t_j be a sequence of numbers. Define $u_j(x^1, x^2) = u(x^1 + t_j, x^2) - u(t_j, 0)$. Then we can find a subsequence of u_j which converges to a generalized solution u_∞ of (4) which is finite everywhere in Ω.*

Lemma 3. *Let u be a solution of (4) and let $\omega_1(t)$ be the oscillation of u in the line segment*

$$\left\{ (x^1, 0) \mid t - \frac{1}{\cos \gamma} \leqslant x^1 \leqslant t + 1 + \frac{1}{\cos \gamma} \right\},$$

$t \in \mathbb{R}$. There exists a constant C_2 such that $\omega_1(t) \leqslant C_2$ for all t.

A proof of the lemma can be found in [13]. In [13], we assumed that $\gamma > 0$, but the proof apparently works for the case $\gamma = 0$ if we only consider the oscillation of u away from the boundary.

Lemma 4. *Let u be a solution of (4) and let $\omega_2(t)$ be the oscillation of u in the line segment $\{(t, x^2) \mid -\frac{1}{2} < x^2 < \frac{1}{2}\}$, $t \in \mathbb{R}$. There exists a constant C_3 such that $\omega_2(t) \leqslant C_3$ for all $t \in \mathbb{R}$.*

PROOF. By Lemma 1, $\omega_2(t)$ is finite for every t. Suppose the lemma is not true, then we can find sequences t_j and x_j^2 with $-\frac{1}{2} < x_j^2 < \frac{1}{2}$ such that

$$\lim_{j \to \infty} |u(t_j, 0) - u(t_j, x_j^2)| = \infty.$$

Using the method in §3.2 in [2], we can prove that

$$u(t_j, x_j^2) \leqslant u(t_j, 0) + \omega_1(t_j) + \frac{1}{\cos \gamma}$$

for all j, where $\omega_1(t_j)$ is the oscillation defined in Lemma 3. Therefore we must have

$$\lim_{j \to \infty} (u(t_j, 0) - u(t_j, x_j^2)) = \infty,$$

by Lemma 3.

Define $u_j(x^1, x^2) = u(x^1 + t_j, x^2) - u(t_j, 0)$. Each u_j is a solution of (4) and

$$\lim_{j \to \infty} u_j(0, x_j^2) = -\infty. \tag{8}$$

By Lemma 2, we can find a subsequence of u_j, also denoted by u_j, which converges locally to a generalized solution u_∞ of (4). Furthermore, u_∞ never takes the value $+\infty$ or $-\infty$ in Ω.

We may also assume that

$$\lim_{j \to \infty} x_j^2 = x_\infty^2 \in \left[-\frac{1}{2}, \frac{1}{2} \right].$$

Denote $x_j = (0, x_j^2)$ and $x_\infty = (0, x_\infty^2)$. Then $\lim_{j \to \infty} x_j = x_\infty \in \bar{\Omega}$.

Let U_j be the subgraphs of u_j and U_∞ be the subgraph of u_∞. Let r_1 and C_1 be the positive constants in Theorem 2. By (8), for any $s > 0$, there is an integer j_s such that

$$u_j(x_j) < -s \qquad \text{for } j \geqslant j_s.$$

Hence $|U'_{j,r}(x_j, -s)| > 0$ for all $j \geqslant j_s$ and $r > 0$. By Theorem 2, we have

$$|U'_{j,r_1}(x_j, -s)| \geqslant C_1 r_1^3$$

for all $j \geqslant j_s$. Since u_j converge locally to u_∞ and x_j converge to x_∞, therefore by letting $j \to \infty$ in the above inequality, we have

$$|U'_{\infty,r_1}(x_\infty, -s)| \geqslant C_1 r_1^3$$

for all $s > 0$. This contradicts the fact that u_∞ is finite everywhere in Ω. The proof of the lemma is completed. $\qquad \square$

Combining Lemma 3 and 4, we have the following.

Corollary 1. *Let u be a solution of (4) and let $\omega(t)$ be the oscillation of u in the set $\{(x^1, x^2) \in \Omega | t < x^1 < t + 1\}$, $t \in \mathbb{R}$. Then there exists a constant C_4 such that $\omega(t) \leqslant C_4$ for all t.*

We are now ready to prove the Main Theorem.

Main Theorem. *Every solution u of (4) is of the form $\Phi_{\beta*} + constant$ for some $|\beta*| < 1$, where $\Phi_{\beta*}$ is defined in (6).*

PROOF. As we mentioned before, we must have $H = 2\cos\gamma$ in order that (4) has a solution. Let $\Omega^+ = \{(x^1, x^2) \in \Omega | x^1 \geq 0\}$ and $\Omega^- = \{(x^1, x^2) \in \Omega | x^1 \leq 0\}$.

By Corollary 1, we can find constants $C_5 > 0$ and $a > 0$ such that

$$|u(x^1, x^2)| \leq C_5 |x^1| \tag{9}$$

for all (x^1, x^2) with $|x^1| > a$. For any $|\beta| < 1$ and $t \in \mathbb{R}$, define

$$S(\beta, t) = \sup_{\substack{(x^1, x^2) \in \Omega \\ x^1 = t}} (\Phi_\beta(x^1, x^2) - u(x^1, x^2))$$

$$\text{and } I(\beta, t) = \inf_{\substack{(x^1, x^2) \in \Omega \\ x^1 = t}} (\Phi_\beta(x^1, x^2) - u(x^1, x^2)).$$

It is easy to see by Lemma 1 that $S(\beta, t)$ and $I(\beta, t)$ are finite. For $\frac{1}{2} < \beta < 1$ and $x^1 > a + (2/\cos\gamma)$ we have

$$\Phi_\beta(x^1, x^2) - u(x^1, x^2)$$

$$\geq -\frac{1}{\sqrt{1 - \beta^2}} \cdot \frac{1}{2\cos\gamma} + \frac{\beta}{\sqrt{1 - \beta^2}} x^1 - C_5 x^1$$

$$\geq -\frac{1}{\sqrt{1 - \beta^2}} \cdot \frac{1}{2\cos\gamma} + \frac{1}{2} \frac{\frac{1}{2}}{\sqrt{1 - \beta^2}} \cdot \frac{2}{\cos\gamma} + \left(\frac{1}{2} \frac{\beta}{\sqrt{1 - \beta^2}} - C_5\right) x^1$$

$$= \left(\frac{1}{2} \frac{\beta}{\sqrt{1 - \beta^2}} - C_5\right) x^1.$$

Since

$$\lim_{\beta \to 1} \left(\frac{1}{2} \frac{\beta}{\sqrt{1 - \beta^2}} - C_5\right) = \infty,$$

therefore we can find $1 > \beta_0' > \frac{1}{2}$ such that

$$\liminf_{k \to \infty} I(\beta, k) > 0 \qquad \text{for } 1 > \beta \geq \beta_0'. \tag{10}$$

Similarly we can find $1 > \beta_0'' > \frac{1}{2}$ such that

$$\limsup_{k \to \infty} S(\beta, k) < 0 \qquad \text{for } -1 < \beta \leq -\beta_0''. \tag{11}$$

Let $\beta_0 = \max(\beta_0', \beta_0'')$, then $0 < \beta_0 < 1$. Equations (10) and (11) will hold for all $\beta_0 \leq \beta < 1$ and $-1 < \beta \leq -\beta_0$, respectively. Let

$$\beta_1^* = \inf\left\{\beta \,\Big|\, \liminf_{k \to \infty} I(\beta, k) > 0\right\}.$$

By (10) and (11), $-\beta_0 \leq \beta_1^* \leq \beta_0$. We claim that

$$|\Phi_{\beta_1^*} - u| \leq C_6 \qquad \text{in } \Omega^+ \tag{12}$$

for some constant C_6.

Since $0 < \beta_0 < 1$, there exists a constant $C_7 > 0$ such that

$$\begin{cases} S(\beta, 0) \leqslant C_7 \\ I(\beta, 0) \geqslant -C_7 \end{cases} \quad \text{for all } |\beta| \leqslant \frac{1 + \beta_0}{2} < 1. \tag{13}$$

Take a sequence $\beta_j \downarrow \beta_1^*$ with $|\beta_j| \leqslant (1 + \beta_0/2)$ such that $\lim \inf_{k \to \infty} I(\beta_j, k) > 0$ for all j. For fixed j, there is an integer k_j such that for all $k \geqslant k_j$

$$I(\beta_j, k) > 0.$$

Combining this with the second inequality in (13) we have

$$\Phi_{\beta_j} + C_7 \geqslant u \text{ on } \{0, k\} \times (-\tfrac{1}{2}, \tfrac{1}{2})$$

for all $k \geqslant k_j$. By Theorem 6 in [2], we have

$$\Phi_{\beta_j} + C_7 \geqslant u \qquad \text{in } \Omega^+.$$

Since $\beta_j \to \beta_1^*$, therefore

$$\Phi_{\beta_1^*} + C_7 \geqslant u \qquad \text{in } \Omega^+. \tag{14}$$

On the other hand, by the definition of β_1^* we have $\lim \inf_{k \to \infty} I(\beta, k) \leqslant 0$ for all $-1 < \beta < \beta_1^*$. For fixed $-1 < \beta < \beta_1^*$ with $|\beta| \leqslant (1 + \beta_0/2)$ and for any $\varepsilon > 0$, we can find a subsequence k_i such that $I(\beta, k_i) < \varepsilon$ for all k_i. By Corollary 1 and the fact that $|\beta| \leqslant (1 + \beta_0/2) < 1$, we can find a constant C_8 such that

$$S(\beta, t) \leqslant I(\beta, t) + C_8$$

for all $t \in \mathbb{R}$. C_8 can be chosen to be independent of β as long as $|\beta| \leqslant (1 + \beta_0)/2$. In particular,

$$S(\beta, k_i) \leqslant I(\beta, k_i) + C_8$$

$$\leqslant \varepsilon + C_8.$$

Combining this with the first inequality in (13) and by Theorem 6 in [2] again, we have

$$\Phi_\beta \leqslant u + C_8 + C_7 + \varepsilon \qquad \text{in } \Omega^+.$$

Since the equality is true for all $-1 < \beta < \beta_1^*$ with $|\beta| \leqslant (1 + \beta_0)/2$ and for all $\varepsilon > 0$, therefore

$$\Phi_{\beta_1^*} \leqslant u + C_7 + C_8 \qquad \text{in } \Omega^+. \tag{15}$$

Combining (14) and (15), we can find a constant C_6 so that (12) holds. Similarly, we can find $-\beta_0 \leqslant \beta_2^* \leqslant \beta_0$ and a constant C_9 such that

$$|\Phi_{\beta_2^*} - u| \leqslant C_9 \qquad \text{in } \Omega^-. \tag{16}$$

We want to show that $\beta_1^* = \beta_2^* = \beta^*$. Suppose $\beta_2^* > \beta_1^*$. Let β be such that $\beta_1^* < \beta < \beta_2^*$. Then

$$\Phi_\beta(x^1, x^2) - \Phi_{\beta_1^*}(x^1, x^2)$$

$$= \frac{1}{\sqrt{1-\beta^2}}\left(-\sqrt{\left(\frac{1}{2\cos\gamma}\right)^2 - (x^2)^2}\right) + \frac{\beta}{\sqrt{1-\beta^2}}x^1$$

$$- \frac{1}{\sqrt{1-(\beta_1^*)^2}}\left(-\sqrt{\left(\frac{1}{2\cos\gamma}\right)^2 - (x^2)^2}\right) - \frac{\beta_1^*}{\sqrt{1-(\beta_1^*)^2}}x^1$$

$$\geq -\frac{1}{\sqrt{1-\beta^2}}\frac{1}{2\cos\gamma} + \left(\frac{\beta}{\sqrt{1-\beta^2}} - \frac{\beta_1^*}{\sqrt{1-(\beta_1^*)^2}}\right)x^1.$$

Since $b/\sqrt{1-b^2}$ is an increasing function of b in $(-1, 1)$, therefore

$$\frac{\beta}{\sqrt{1-\beta^2}} - \frac{\beta_1}{\sqrt{1-(\beta_1^*)^2}} > 0 \quad \text{and} \quad \lim_{x^1 \to \infty} (\Phi_\beta(x^1, x^2) - \Phi_{\beta_1^*}(x^1, x^2)) = \infty.$$

Similarly we can prove that

$$\lim_{x^1 \to -\infty} (\Phi_\beta(x^1, x^2) - \Phi_{\beta_1^*}(x^1, x^2)) = \infty.$$

By (12) and (16), therefore we have

$$\lim_{|x^1| \to \infty} [\Phi_\beta(x^1, x^2) - (u(x^1, x^2) + C)] = \infty$$

for all $C > 0$. By Theorem 6 of [2], we have $\Phi_\beta \geq u + C$ in Ω for all $C > 0$, which is impossible. Hence, we must have $\beta_2^* \leq \beta_1^*$.

Similarly, we can prove that $\beta_1^* \leq \beta_2^*$. Therefore $\beta_1^* = \beta_2^* = \beta^*$ and

$$|\Phi_{\beta^*} - u| \leq C_{10} \quad \text{in } \Omega \tag{17}$$

where $C_{10} = \max(C_6, C_9)$.

The following inequality is observed by Wong [15]:

$$(p - q) \cdot \left(\frac{p}{\sqrt{1+|p|^2}} - \frac{q}{\sqrt{1+|q|^2}}\right) \geq \left|\frac{p}{\sqrt{1+|p|^2}} - \frac{q}{\sqrt{1+|q|^2}}\right|^2 \tag{18}$$

for any $p, q \in \mathbb{R}^n$.

We will use (18) in order to simplify the notation, but one can prove the theorem in essentially the same way without using (18). In order to simplify the notation further, let us denote $v = \Phi_{\beta^*}$.

For $t > 0$, we have

$$\int_{\Omega_t} (v - u)\operatorname{div}(Tv - Tu)\,dx = 0.$$

Integrating by parts and using (17) and (18),

$$\int_{\Omega_t} |Tv - Tu|^2 \, dx \leqslant \int_{\Omega_t} (Dv - Du) \cdot (Tv - Tu) \, dx$$

$$= \int_{\{-t,t\} \times (-1/2, 1/2)} (v - u)(Tv - Tu) \cdot v \, dH_1$$

$$\leqslant C_{10} \int_{\{-t,t\} \times (-1/2, 1/2)} |Tv - Tu| \, dH_1, \qquad (19)$$

where H_1 is the one-dimensional Hausdorff measure.

Define $f(t) = \int_{\Omega_t} |Tv - Tu|^2 \, dx$. By (19) and using Schwarz's inequality, we have

$$f(t) \leqslant C_{10} \sqrt{2} \left(\int_{\{-t,t\} \times (-1/2, 1/2)} |Tv - Tu|^2 \, dH_1 \right)^{1/2}$$

$$= C_{10} \sqrt{2} (f'(t))^{1/2}.$$

If $f(t_0) > 0$ for some $t_0 > 0$, then $f(t) > 0$ for all $t \geqslant t_0$ and

$$\frac{f'(t)}{(f(t))^2} \geqslant \frac{1}{2C_{10}^2}$$

for all $t \geqslant t_0$. Integrate the above inequality from t_0 to t, $t > t_0$

$$\frac{1}{2C_{10}^2} (t - t_0) \leqslant \frac{1}{f(t_0)} - \frac{1}{f(t)} \leqslant \frac{1}{f(t_0)}.$$

Let $t \to \infty$, we have a contradiction.

Hence $f(t) \equiv 0$ and $Du \equiv Dv$. Therefore $u = v + \text{constant} = \Phi_{\beta*} + \text{constant}$. This completes the proof of the theorem.

References

[1] S.N. Bernstein, *Sur les surfaces definies au moyen de leur courbure moyenne ou totale*, Ann. Ec. Norm. Sup. **27** (1908), 233–256.

[2] P. Concus and R. Finn, *On capillary free surfaces in the absence of gravity.* Acta Math. **132** (1974), 117–198.

[3] R. Finn, *Remarks relevant to minimal surfaces and to surfaces of prescribed mean curvature*, Journ. d'Anal. Math. **14** (1965), 139–160.

[4] R. Finn, *A subsidiary variational problem and existence criteria for capillary surfaces*, Indiana Univ. Math. J. **32** (1983), 439–460.

[5] D. Gilbarg and N.S. Trudinger, *Elliptic partial differential equations of second order.* Springer-Verlag, Berlin-Heidelberg-New York, 1977.

[6] E. Giusti, *Minimal surfaces and functions of bounded variation.* Notes on pure mathematics, Australian National Univ., Canberra, 1977.

[7] E. Giusti, *On the equation of surfaces of prescribed mean curvature. Existence and uniqueness without boundary conditions*, Inv. Math. **46** (1978), 111–137.

[8] E. Giusti, *Generalized solutions of mean curvature equations*, Pac. J. Math. **88** (1980), 239–321.

[9] U. Massari, *Frontiere orientate di curvature media assegneta in L^p*, Rend. Sem. Mat. Padova **53** (1975), 37–52.

[10] U. Massari and L. Pepe, *Sulle successioni convergenti di superfici a curvatura media assegnata*, Rend. Sem. Mat. Padova **53** (1975), 53–68.

[11] M. Miranda, *Superfici minime illimitte*, Ann. Scuola Norm. Sup. Pisa, (4) **4** (1977), 313–322.

[12] L.-F. Tam, *On existence criteria for capillary free surfaces without gravity*, Pac. J. Math. (to appear).

[13] L.-F. Tam, *On the uniqueness of capillary surfaces without gravity over an infinite strip*, Indiana Univ. Math. J. (to appear).

[14] L.-F. Tam, *The behavior of capillary surfaces as gravity tends to zero*, Comm. in P.D.E. (to appear).

[15] C. Wong, Oral communication.

The Behavior of a Capillary Surface for Small Bond Number

David Siegel

1. Introduction

The boundary value problem

$$\text{div}(Tu) = \kappa u \qquad \text{in } \Omega$$
$$Tu \cdot v = \cos \gamma \qquad \text{on } \Sigma = \partial \Omega \tag{1}$$

determines the height $u(x)$ of a capillary surface. Here κ is a positive constant, Ω is a bounded domain in R^n, v is the exterior normal on Σ, and Tu is the vector operator

$$Tu = \frac{\nabla u}{\sqrt{1 + |\nabla u|^2}}.$$

For $n = 2$, $u(x)$ can be interpreted physically as the height of a capillary surface above the undisturbed fluid level when a vertical cylindrical tube with section Ω is dipped into a large reservoir. The capillarity constant $\kappa = \rho g/\sigma$ where ρ is the density difference across the surface, g the gravitational acceleration, σ the surface tension. The contact angle γ is constant if the tube is homogeneous, and $0 \leq \gamma \leq \pi$.

It is convenient to write the equation in nondimensional form [1]. Let "a" be a representative length, set $U = u/a$ and $X = x/a$, then the equation and boundary condition become

$$\text{div}(TU) = BU \qquad \text{in } \Omega_0 = \frac{1}{a}\Omega$$
$$TU \cdot v = \cos \gamma \qquad \text{on } \Sigma_0 = \frac{1}{a}\Sigma, \tag{2}$$

where B is the nondimensional "Bond number", $B = \kappa a^2$. Among other possibilities, a small bond number corresponds to a narrow tube or a low force of gravity.

For $\Omega_0 = B_1(0) = \{x : |x| < 1\}$, the ball of radius 1 centered at the origin, the solution $U(r; B, \gamma)$ to (2) satisfies an ordinary differential equation in terms of r, $r = |x|$. This solution has been studied in [2]–[6]. Theorem 8 from [3] implies that

$$U(r; B, \gamma) = \frac{n \cos \gamma}{B} + H(r; B, \gamma) + 0(B) \tag{3}$$

as B tends to zero, where H is the spherical cap such that $\int_{\Omega_0} H \, dx = 0$ and $TH \cdot v = \cos \gamma$ on Σ_0.

In this paper we extend the preceding result to a general domain Ω_0 with a sufficiently smooth boundary. Our main assumption is that the zero gravity capillary surface $V(x; \gamma)$ exists and is C^1 on the closed domain $\bar{\Omega}_0$. V is determined by

$$
\begin{aligned}
NV &= A &&\text{in } \Omega_0 \\
TV \cdot v &= \cos \gamma &&\text{on } \Sigma_0 \\
\int_{\Omega_0} V dx &= 0.
\end{aligned}
\tag{4}
$$

Here $A = \cos \gamma |\Sigma_0|/|\Omega_0|$, with $|\ |$ denoting the appropriate measure, and $NV = \text{div}(TV)$. The existence of such surfaces is discussed in [7] and [8]. Under the additional assumptions that Ω_0 is a $C^{2,\alpha}$ domain and $0 < \gamma < \pi$, we will prove

$$U(x; B, \gamma) = \frac{A}{B} + V(x; \gamma) + 0(B) \tag{5}$$

as B tends to zero.

A related result of Emmer [9] shows that $\lim_{B \to 0} U(x; B, \gamma) = \infty$ in $\bar{\Omega}_0$ under the less restrictive hypotheses that Σ_0 is Lipschitz with Lipschitz constant L, and $0 < \cos \gamma < 1/\sqrt{1 + L^2}$. Emmer does not assume the existence of V.

2. Proof of Result

Let $U = A/B + W$, then W satisfies

$$
\begin{aligned}
NW &= A + BW &&\text{in } \Omega_0 \\
TW \cdot v &= \cos \gamma &&\text{on } \Sigma_0
\end{aligned}
\tag{6}
$$

Lemma. $-\max V < W - V < -\min V$.

PROOF. We note that $N(V - \min V) = A \leq A + B(V - \min V)$ in Ω_0 and $T(V - \min V) \cdot v = \cos \gamma$ on Σ_0. By the comparison principle of Concus and Finn [10] we obtain $W < V - \min V$. By a similar argument we obtain $W > V - \max V$.

In what follows, C_i will denote constants that do not depend upon B, for $B \leq B_0$, B_0 a fixed arbitrary value. The lemma implies

$$|W| \leq C_1. \tag{7}$$

From Theorem 1.1 of [11] we obtain

$$|\nabla W| \leq C_2. \tag{8}$$

An application of the divergence theorem gives

$$\int_{\Omega_0} (\nabla W - \nabla V) \cdot (TW - TV) \, dx = -\int_{\Omega_0} (W - V)(NW - NV) \, dx$$

$$= -B \int_{\Omega_0} (W - V) W \, dx. \tag{9}$$

Let $A^i(P) = P_i / \sqrt{1 + P^2}$ and $u_t = tW + (1 - t)V$, then $A^i(\nabla W) - A^i(\nabla V) = a^{ij} D_j(W - V)$ with

$$a^{ij} = \int_0^1 D_{P_j} A^i(\nabla u_t) \, dt.$$

Now

$$D_{P_j} A^i = \frac{\delta_{ij}}{\sqrt{1 + |P|^2}} - \frac{P_i P_j}{(1 + |P|^2)^{3/2}}$$

and the well known inequality

$$D_{P_j} A^i \xi_i \xi_j \geq \frac{\xi^2}{(1 + |P|^2)^{3/2}}$$

imply

$$(\nabla W - \nabla V) \cdot (TW - TV) \geq |\nabla W - \nabla V|^2 \int_0^1 \frac{dt}{(1 + |\nabla u_t|^2)^{3/2}}. \tag{10}$$

Combining (7)–(10) we obtain

$$\int_{\Omega_0} |\nabla W - \nabla V|^2 \, dx \leq C_3 B \int_{\Omega_0} |W - V| \, dx. \tag{11}$$

By integrating (6a) over Ω_0 and using the divergence theorem we find $\int_{\Omega_0} W \, dx = 0$, so that $\int_{\Omega_0} (W - V) \, dx = 0$.

We can apply the Poincaré inequality, which holds for piecewise smooth domains (see (2.10) in [12]), to $W - V$ and obtain

$$\int_{\Omega_0} |W - V|^2 \, dx \leq C_4 \int_{\Omega_0} |\nabla W - \nabla V|^2 \, dx. \tag{12}$$

Combining (11) and (12) and applying the Cauchy–Schwarz inequality we obtain

$$\|W - V\|_2 = \left(\int_{\Omega_0} |W - V|^2 \right)^{1/2} \leq C_5 B. \tag{13}$$

Finally we obtain a pointwise estimate. Let $Z = W - V$, then Z satisfies

$$\text{div}(a^{ij} D_i Z) = d = BW \qquad \text{in } \Omega_0$$

$$a^{ij} D_i Z v^j = 0 \qquad \text{on } \Sigma_0.$$

A result of Lieberman [13, Lemma 3.1] gives

$$\sup |Z| \leq C_6 \left[\int_{\Omega_0} |Z| \, dx + \sup |d| \right],$$

thus

$$\sup |W - V| \leq C_7 B. \tag{14}$$

Using the definition of W, we see that (14) implies the conclusion (5). $\qquad \square$

3. Remarks and Open Questions

(a) If the contact angle γ is not constant $0 < \gamma(\sigma) < \pi$, with γ a $C^{1,\beta}$ function, then the same result (5) is true provided we take A to be $\int_{\Sigma_0} \cos \gamma(\sigma) \, d\sigma / \Omega_0$.

(b) If $\gamma = 0$, the above argument shows only that $u = A/B + 0(1)$ as $B \to 0$. We expect that (5) is true since it is true (Theorem 8 of [3]) for $\Omega_0 = B_1(0)$.

(c) The most intriguing question is what happens when V does not exist. One would expect some sort of singular behavior.

Acknowledgment. The author wishes to thank Gary Lieberman for pointing out the applicability of his result.

References

1. R. Finn, *Equilibrium Capillary Surfaces*, Springer-Verlag (to appear).
2. R. Finn, *Some comparison properties and bounds for capillary surfaces*, Complex Analysis and its Applications, Moscow Math. Soc., volume dedicated to I.N. Vekua, Scientific Press, Moscow, 1978.
3. D. Siegel, *Height estimates for capillary surfaces*, Pacific J. Math., **88** (1980), 471–515.
4. R. Finn, *On the Laplace formula and the meniscus height for a capillary surface*, ZAMM **61** (1981), 165–173.

5. R. Finn, *Addenda to my paper "On the Laplace formula..."*, ZAMM **61** (1981), 175–177.

6. F. Brulois, *Arbitrarily precise estimates for rotationally symmetric capillary surfaces*, Dissertation, Stanford University, 1981.

7. Jin-Tzu Chen, *On the existence of capillary free surfaces in the absence of gravity*, Pac. J. Math., **88** (1980), 323–361.

8. R. Finn, *Existence criteria for capillary free surfaces without gravity*, Indiana Math J., **32** (1983), No. 3, 439–460.

9. M. Emmer, *On the behavior of the surface of equilibrium in the capillary tubes when gravity goes to zero*, Rend. Sem. Mat. Univ. Padova, **65** (1981), 143–162.

10. P. Concus and R. Finn, *On capillary free surfaces in a gravitational field*, Acta Math., **132** (1974), 207–223.

11. C. Gerhardt, *Global regularity of the solutions to the capillarity problem*, Ann. Scoula Nor. Sup., Pisa, Classe di Scienze, Series IV, Vol. III, n. 1 (1976), 157–175.

12. D.A. Ladyzhenskaya and N.N. Ural'tseva, *Linear and Quasilinear Elliptic Equations*, Academic Press, New York, 1966.

13. G. Lieberman, *The conormal derivative problem for elliptic equations of variational type*, JDE, **49** (1983), 218–257.

Convexity Properties of Solutions to Elliptic P.D.E.'S

Nicholas J. Korevaar

How do the data of an elliptic boundary value problem (domain, boundary values, elliptic operator) affect the shape of the solution v? Estimates for v, Dv, or even D^2v may be necessary to prove existence and regularity theorems, and they often also characterize fundamental geometric behavior of v. In this note we shall study some particular estimates involving v, Dv, D^2v: ones that are related to convexity properties of v. The results do not usually lead to existence theorems (with some exceptions, e.g. [3]), but are surprising and have independent beauty.

Results

The following four examples show the type of results one can prove about solutions to semilinear problems with constant Dirichlet data. (Domains Ω are $C^{2,\alpha}$ and compact in this note. Ω is called strictly convex here if at every point of $\partial\Omega$ all principal curvatures are strictly positive. Similarly u is called strictly convex if $D^2u > 0$.) Note that in most cases it is not the solution v to the original problem that is convex, but rather some $u = g(v)$, with $g' > 0$. This still yields information about v: The level sets $\{v \leqslant c\} = \{u \leqslant g^{-1}(c)\}$ are convex, v has a unique minimum, and D^2v can be estimated cleanly in terms of v and Dv (by translating "$D^2u > 0$" into a statement about D^2v).

P1 $\begin{cases} \Delta v = 1 & \text{in } \Omega \text{ convex} \\ v = 0 & \text{on } \partial\Omega \end{cases}$ $u = -(-v)^{1/2}$ is strictly convex

P2 $\begin{cases} \Delta v = -\lambda v, v < 0 & \text{in } \Omega \text{ convex} \\ v = 0 & \text{on } \partial\Omega \end{cases}$ $u = -\ln(-v)$ is strictly convex

$$P3 \begin{cases} \Delta v = 0 & \text{in } \Omega_1 \backslash \Omega_2, \Omega_2 \subset\subset \Omega_1, \\ v = 0 & \text{on } \partial\Omega_1 \\ v = -1 & \text{on } \partial\Omega_2 \end{cases} \qquad \begin{array}{l} u = (v + M)^N \text{ is strictly convex} \\ \text{for } M, N(M) \text{ large} \end{array}$$

$$P4 \begin{cases} \Delta v = e^v & \text{in } \Omega \subset \mathbb{R}^2, \text{ convex} \\ v \to \infty & \text{as } x \to \partial\Omega \end{cases} \qquad u = v \text{ is strictly convex}$$

Method

The examples above may be viewed as consequences of a "structure" theorem proved in $\mathbb{R}^2$ by L. Caffarelli and A. Friedman [2] and generalized to $\mathbb{R}^n$ by this author and J. Lewis [11]. In fact, this theorem seems to yield all the currently known results about convexity of level sets, which no one technique could do before.

The convexity results of P1–P4 were known before our work (that of J. Lewis and this author), but the strict convexity is a new consequence of the structure theorem. R. Gabriel first proved that the level sets of v in P3 are convex (hence also those of Green's functions in convex domains), using a functional that worked directly with the level sets [4] (or see [7]). H. Brascamp and E. Lieb first showed the result of P2, using the related heat equation and integral properties of logconcave functions [1].

The maximum principle approach for studying convexity of solutions to elliptic problems yielded results (in capillarity) when a maximum principle was discovered for the "concavity" functional

$$\mathbb{C}(x, z) = u\left(\frac{x + z}{2}\right) - \frac{1}{2}u(x) - \frac{1}{2}u(z)$$

(valid for solutions u to a class of elliptic operators) [9]. (Note $\mathbb{C} \leqslant 0 \; \forall (x, z) \in \Omega \times \Omega$ iff u convex.) This was applied to P2 in [10], and by Caffarelli and J. Spruck in [3]. A. Kennington and (to an intermediate degree) B. Kawohl generalized the concavity maximum principle. The full generalization was applied to P1, P4, and other problems [8], [5], [6].

A good current summary of what is known about convexity and semilinear boundary value problems, including a large bibliography, is the set of lecture notes by Kawohl, "Rearrangements and convexity of level sets in P.D.E." [7].

The structure theorem we consider is:

Theorem. *Let $u \in C^4(\Omega)$ satisfy $\Delta u = f(u, Du)$ and be convex, $[D^2u] \geqslant 0$. If*

$$f > 0, (1/f)_{uu} \geqslant 0, \tag{1}$$

then the rank r of D^2u is constant in Ω, $1 \leqslant r \leqslant n$. If $r < n$ the graph of u is a ruled surface. If also $(1/f)_{uu} > 0$, then the graph of u is a cylinder and in fact u is constant in $(n - r)$ fixed directions.

Remarks. An analogous theorem holds for convex solutions to $a^{ij}(Du)u_{ij} = f(x, u, Du)$ if $1/f(x, u, p)$ is convex in (x, u) for fixed p (see [11]). An important consequence of the structure theorem is that if $D^2 u > 0$ at one point of Ω, then $D^2 u > 0$ throughout Ω.

The structure theorem can be used to prove convexity theorems via a homotopy method, developed by Caffarelli–Friedman in [2]. (Earlier, S.T. Yau used a similar method to give a specialized proof of P3 [13].) Suppose $u = g(v)$, where v satisfies a semilinear boundary value problem and suppose that A, B, C hold:

A. u is a strictly convex (radial) function for $\Omega_0 = B_1 = \{|x| < 1\}$, or in ring problems like P3, for $\Omega_0 = B_1 \backslash B_{1/2}$.
B. $\exists$ a homotopy of smooth, strictly convex domains $\Omega(t)$ with $\Omega(0) = \Omega_0$, $\Omega(1) = \Omega$ such that the solutions $v(t, x(t))$ (hence $u(t, x(t))$) vary in a C^2 manner as t varies: $\forall K \subset\subset \Omega_0$, $0 < t_1 < t_2 \leqslant 1$, we have $|v(t_1, x(t_1)) - v(t_2, x(t_2))|_{C^2(K)} \to 0$ as $t_1 \to t_2$.
C. $[D_x^2 u](t, x(t)) > 0 \ \forall x(t)$ sufficiently near $\partial\Omega(t)$ (depending on t).

Then the function u satisfies $D^2 u > 0$ in Ω: if not, there would be some $0 < t \leqslant 1$ for which the corresponding u was convex but for which $D^2 u$ was not everywhere positive, because of A, B, C. But C combined with the structure theorem would then imply a contradiction.

For a given homotopy $\Omega(t)$ the smooth dependence of v on t will depend on some type of elliptic regularity theory. For example, if $\Delta v = h(v)$ with $h_v \geqslant 0$, one can use a version of the general implicit function theorem. In general it may be harder.

Accepting the smooth dependence of v on t it remains to understand the choice of transformation $v = g(u)$ that leads to an equation $\Delta u = f(u, Du)$ satisfying (1), A, C. For $\Delta v = h(v, Dv)$ we get

$$u_i = g'v_i$$

$$u_{ij} = g''v_iv_j + g'v_{ij} \tag{2}$$

$$\Delta u = g''(g')^{-2}|Du|^2 + g'h = f(u, Du).$$

If $h \geqslant 0$, $v|_{\partial\Omega} \equiv 0$, then $v_n < 0$ on $\partial\Omega$, by the strong maximum principle (n = inner normal). The strict convexity of a domain Ω (we suppress t) and an easy computation imply $[v_{ij}]$ is positive in tangential directions on $\partial\Omega$. But $[v_iv_j]$ is positive in any non-tangential direction since Dv is parallel to n. Thus if $|D^2v| < M$ and g'' is large compared to Mg' near $\partial\Omega$, it follows from (2) that $[u_{ij}] > 0$ in a neighborhood of $\partial\Omega$. (One way to insure this is to have g', $g''/g' \to +\infty$ as $v \to 0$. The effect of such a transformation is to bend the graph of u, making it nearly vertical near $\partial\Omega$.)

It is more subtle to pick g so that (1) is satisfied for $f(u, Du)$ in (2). (Note $f > 0$ holds if h, g', $g'' > 0$.) If one makes the Ansatz (see [2], [6]) that $f(u, Du)$ have the form $k(u)(|Du|^2 + 2)$ (the 2 is arbitrary) then $(1/f)_{uu}$ is simpler to compute. The Ansatz yields

$$g' = \frac{1}{\sqrt{-H}}, \qquad H(v) = \int_0^v h(s)\, ds.$$

For a class of equations, this transformation leads to suitable f (one translates $2f_u^2 - ff_{uu} \geqslant 0$ to conditions on h, using the chain rule several times). In any case, it is this choice of g' that yields the correct transformations for P1, P2, and for some other real problems. (Note g', $g''/g' \to \infty$ as $v \to 0$.) There is no known algorithm for finding g given arbitrary h, however.

Remark. There is a connection between the structure theorem and the concavity maximum principle (CMP) mentioned earlier. In order for $\mathbb{C}$ to not have a positive interior maximum $(x, z \in \Omega)$, one needs that $f_u > 0$, in addition to (1). If this additional hypothesis holds, one can work directly on Ω, which is easier than homotopying. In P1 and P2 this approach was the first one to be used. In contrast, problem P3 and its generalizations must be attacked via homotopy, because $f_u < 0$ there.

Structure Theorem Proof

The structure theorem can be thought of as a microscopic version of the CMP and (for real analytic f) a proof can be given which is very similar to the CMP proof in [7]. The general proof follows different lines, however (motivated by [2]). (Both proofs are given in [11], along with the results about ruled surfaces and cylinders, which we omit here.)

To prove the theorem we show that for any $1 \leqslant s \leqslant n$ the set

$$\Omega_s = \{x \in \Omega \text{ s.t. rank } D^2 u(x) \leqslant s\}$$

is open and closed. Hence rank $D^2 u \equiv r$, where r is the smallest rank for which Ω_r is nonempty.

For any $1 \leqslant s \leqslant n$ define ϕ^s to be a function of symmetric, non-negative matrices, the sum of all principal $(s+1) \times (s+1)$ minors:

$$\phi^s(A) = \sum_{\substack{i_1 < \cdots < i_{s+1} \\ \sigma \in S_{s+1}}} (-1)^\sigma A_{i_1 i_{\sigma(1)}} \cdots A_{i_{s+1} i_{\sigma(s+1)}} = \sum_{i_1 < \cdots < i_{s+1}} \lambda_{i_1} \cdots \lambda_{i_{s+1}}, \qquad (3)$$

where the λ_k are the eigenvalues of A. Since Ω_s is the zero set of $\phi^s(D^2 u(x))$, it is closed. The content of the structure theorem is that Ω_s is open.

Let $x \in \Omega_s$. Assume rank $D^2 u(x) = s$. (If the rank is $s' < s$ then repeat the argument below with s' in place of s.) By the strong maximum principle for elliptic equations it suffices to show that $\exists$ a neighborhood $x \in U$ and a constant C so that $\forall z \in U$,

$$\Delta \phi \leqslant C \cdot (\phi + |D\phi|), \qquad (4)$$

since $\phi \geqslant 0$, $\phi(x) = 0$ then implies $\phi \equiv 0$ in U. (We write ϕ for $\phi^s(D^2 u(z))$.) The proof is technical but to facilitate later comments we sketch its key points: If

$A(y)$ is a smooth, non-negative, symmetric matrix valued function of y with rank $A(x) = s$, then $\exists \delta > 0$ s.t. for any fixed z near enough x we may (after orthogonal conjugation of A) assume A has the form

$$
\begin{bmatrix}
A_{11} & & & & & & \\
 & A_{22} & & & & 0 & \\
 & & \ddots & & & & \\
 & & & A_{ss} & & & \\
 & 0 & & & \ddots & & \\
 & & & & & & A_{nn}
\end{bmatrix},
\tag{5}
$$

where $A^{11}, \dots, A_{ss} > \delta$. Let j's index "good" numbers between 1 and s, $j \in G$. Let i's index "bad" numbers from $s + 1$ to n, $i \in B$. Using the formula (3) for ϕ (and its invariance with respect to orthogonal conjugation), and the choice (5) of A at z, one can compute that for

$$
Q = \prod_{j \in G} A_{jj} \qquad Q_j = \frac{Q}{A_{jj}} \qquad R = \sum_{j \in G} Q_j
$$

and for a sufficiently small neighborhood U of x, $\exists C = C(U, \delta, |A|_{C^2})$ so that if $z \in U$,

$$
\Delta\phi(A(z)) \leqslant Q \left[\sum_{i \in B} \Delta A_{ii}(z) - 2 \sum_{\substack{j \in G \\ i \in B}} \frac{|DA_{ij}|^2}{A_{jj}} \right] + C \cdot (\phi + |D\phi|).
\tag{6}
$$

In the case $A = D^2 u$ one uses the equation $\Delta u = f(u, Du)$, its first two derivatives, and a Cauchy–Schwartz estimate to conclude

$$
\Delta\phi \leqslant Q \left(f_{uu} - 2 \frac{f_u^2}{f} \right) \sum_{i \in B} u_i^2 + C \cdot (\phi + |D\phi|).
\tag{7}
$$

By the hypotheses on f this suffices, since $f_{uu} - 2f_u^2/f \leqslant 0$! Thus ϕ is zero in a neighborhood of x and Ω_s is open. $\qquad \square$

Remark. From this proof one cannot be sure that the condition $(1/f)_{uu} \geqslant 0$ is sufficient or natural. The following example shows it is both. Let f be given with $f > 0$, $(1/f)_{uu} < 0$ for some argument (u, p). By translation, rotation, and scaling assume $(u, p) = (0, e_1)$ (where e_1 is the first coordinate vector). We construct an example in two variables which extends trivially to one in n variables. Assume f is real analytic. Use Cauchy–Kowaleski to solve, near the origin, the Cauchy problem

$$
\Delta u = f(u, Du)
$$

$$
u(x, 0) = x + \varepsilon x^4
$$

$$
u_y(x, 0) = 0.
$$

It is a straightforward (but nontrivial) computation, using the power series expansion of u, that for $\varepsilon > 0$ sufficiently small (depending on how much $(1/f)_{uu} < 0$) there is a punctured neighborhood of 0 on which $u_{yy} > 0$ and $\det(D^2u) > 0$. Hence rank D^2u is 1 at the origin but 2 in a punctured neighborhood of it.

Remark. The arguments of the structure theorem are local. Thus one might hope to mimic the proof and derive an analogous theorem on general Riemannian manifolds: Consider the matrix of the second covariant derivatives of u (with respect to an orthonormal frame) and try to prove a rank theorem. Between (6) and (7) one must interchange order of differentiation, so that curvature terms enter. A. Treibergs and this author completed such a calculation. On a manifold with non-negative constant curvature K: $R_{ijkl} = K(\delta_{ik}\delta_{jl} - \delta_{il}\delta_{jk})$ (that is, space, spheres, or their quotients), inequality (7) becomes

$$\Delta\phi \leq Q\left[\left(f_{uu} - 2\frac{(f_u + Ks)^2}{f}\right)\sum_{i \in B} u_i^2\right.$$
$$\left. - K\left(2f(n - s) - \sum_{j \in G} f_{pj}u_j(n - s) - \sum_{i \in B} f_{pi}u_i(n - s - 1)\right)\right]$$
$$+ C(\phi + |D\phi|). \tag{8}$$

For f of the form $M(u) + N(u)|Du|^2$ with $M, N \geq 0$, the second collection of terms in (8) is non-negative and so one gets a nice extension of (7):

$$\Delta\phi \leq Q\left[f_{uu} - 2\frac{(f_u + Ks)^2}{f}\right]\sum_{i \in B} u_i^2. \tag{9}$$

One can then, for example, use the corresponding structure theorem and homotopy method to prove analogs of P1 and P2 for these spaces.

Unfortunately, if the manifold's curvature is not constant, the terms do not group so fortuitously; we have been unable to extend any version of structure theorem beyond constant curvature spaces.

Questions

What type of structure theorem can hold for convex solutions to semilinear equations on Riemannian manifolds (depending on curvature conditions)? Specifically, consider the simple manifold $S^2 \times S^2$. Does the first eigenfunction of the Laplacian on a convex domain have convex level sets?

If v satisfies any reasonable semilinear equation, $\Delta v = h(v, Dv), h > 0$, with constant Dirichlet data, does it have a unique minimum? Are its level sets convex? Given h, is it possible to find, in a natural (or any) way, a transformation $u = g(v)$ so that the structure theorem applies to u? (The cases where this method works are relatively rare so far.)

Once one knows $D^2u > 0$, can one get an estimate $D^2u > \delta$, with δ depending somehow on the data of the semilinear equation? Such estimates might be useful in trying to prove existence and regularity results.

Are there other properties of the boundary, besides convexity, that are reflected by solutions to appropriate boundary value problems? (Starshapedness is one example [7].)

For data other than constant Dirichlet values (e.g. constant Neumann or constant contact angle data), and/or for operators other than the Laplacian (e.g. mean curvature), what can be said about the level sets of the solutions?

Acknowledgment. Research was supported by National Science Foundation Grant No. DMS 85-11478.

References

[1] H.J. Brascamp and E.H. Lieb, *On extensions of the Brunn–Minkowski and Prekopa–Leindler theorems, including inequalities for log concave functions, and with an application to the diffusion equation,* J. Fcnl. Anal. **22** (1976), 366–389.

[2] L.A. Caffarelli and A. Friedman, *Convexity of solutions of semilinear elliptic equations,* Duke M. Jour. **52** #2 (1985), 431–456.

[3] L.A. Caffarelli and J. Spruck, *Convexity properties of solutions to some classical variational problems,* Comm. Part. Diff. Eq. **7** (1982), 1337–1379.

[4] R. Gabriel, *A result concerning convex level surfaces of 3-dimensional harmonic functions,* J. London Math. Soc. **32** (1957), 286–294.

[5] B. Kawohl, *A remark on N. Korevaar's concavity maximum principle and on the asymptotic uniqueness of solutions to the plasma problem,* Math. Meth. Appl. Sci. **8** (1986), 93–101.

[6] B. Kawohl, *When are solutions to nonlinear elliptic boundary value problems convex?,* Comm Part. Diff. Eq. (to appear).

[7] B. Kawohl, *Rearrangements and convexity of level sets in P.D.E.,* Lecture Notes in Mathematics, Springer-Verlag, New York, 1985.

[8] A. Kennington, *Power concavity and boundary value problems,* Ind. U. Math. J. **34** #3 (1985), 687–704.

[9] N.J. Korevaar, *Capillary surface convexity above convex domains,* Ind. U. Math. J. **32** #1 (1983), 73–81.

[10] N.J. Korevaar, *Convex solutions to nonlinear elliptic and parabolic boundary value problems,* Ind. U. Math. J. **32** (1983), 603–614.

[11] N.J. Korevaar and J.L. Lewis, *Convex solutions to certain elliptic P.D.E.'s have constant rank Hessians,* Arch. Rat. Mech. Anal. (to appear).

[12] J.L. Lewis, *Capacitary functions in convex rings,* Arch. Rat. Mech. Anal. **66** (1977), 201–224.

[13] A. Treibergs, personal communication (1984).

Boundary Behavior of Capillary Surfaces Via the Maximum Principle

Gary M. Lieberman

Various authors have studied the boundary behavior of capillary surfaces using deep techniques. This talk is concerned with obtaining the same results via the maximum principle.

Specifically we consider non-parametric capillary surfaces. Thus we examine solutions u of the nonlinear boundary value problem

$$\operatorname{div} v = \kappa u \quad \text{in } \Omega, \qquad v \cdot \gamma + \cos \phi = 0 \quad \text{on } \partial\Omega \tag{1}$$

for constants $\kappa > 0$ and $0 \leqslant \phi \leqslant \pi$, where $v = (1 + |Du|^2)^{-1/2}|Du|$ and γ is the inner normal to $\partial\Omega$. We always assume that u and Ω are at least smooth enough that the equations of (1) hold everywhere and that the derivatives entering into these equations are continuous. Under these circumstances, the study of boundary behavior is reduced to certain *a priori* estimates on u and its derivatives; well-known approximation arguments can be used to conclude that these estimates (and the implied regularity results) apply to variational solutions of (1) when $\partial\Omega$ is smooth enough. More generally, many of the results mentioned here are valid for the problems

$$\operatorname{div} v = H(x, u) \quad \text{in } \Omega, \qquad v \cdot \gamma + \cos \theta(x) = 0 \quad \text{on } \partial\Omega, \tag{1$'$}$$

where H and θ are smooth with $H_z \geqslant 0$ (here and below z and p are used as replacement variables for u and Du, respectively), and

$$a^{ij}(x, u, Du)D_{ij}u + a(x, u, Du) = 0 \quad \text{in } \Omega, \qquad b(x, u, Du) = 0 \quad \text{on } \partial\Omega, \tag{2}$$

where the matrix (a^{ij}) is positive definite and the directional derivative $b_p \cdot \gamma$ is positive; further hypotheses are needed on the functions a^{ij}, a, and b, but we shall not spell them out here.

The first estimate is a bound on the maximum absolute value of u. Such a bound was proved for problem (1) by Ural'tseva [11] using test function arguments; Concus and Finn [1] used a simple variant of the maximum principle for this bound. They assume that

$$(\operatorname{sgn} z)H(\cdot, z) \to \infty \qquad \text{as } z \to \infty$$

uniformly in Ω, in which case lower (upper) hemispheres are subsolutions (supersolutions) of the differential equation after suitable translation; moreover, the verticality of the hemisphere at its boundary precludes a boundary minimum for the difference between the solution and the subsolution. When Ω satisfies an interior sphere condition (for example, if the domain has C^2 boundary), it follows that the maximum absolute value of u can be estimated without recourse to the boundary condition. Otherwise this estimate follows if the boundary and contact angle satisfy an appropriate relation by a more careful analysis of the boundary condition. This approach has been adapted to problem (2) by the present author in [6, Sect. 3].

The next estimate is a bound on the size of Du, first obtained by Ural'tseva [12] by test function arguments. It is clear that this bound requires $|\cos \theta|$ to be bounded away from 1. Two related maximum principle approaches to this bound have been developed [4,8]. The first approach, due to Korevaar, uses a normal perturbation similar to the one described in [3] for interior gradient bounds for the minimal surface equation. Here the idea is to create a comparison function v by moving the graph of u first a distance $\varepsilon f(x, u(x))$ along its normal vector and then a distance $\varepsilon g(x, u(x))$ along a suitable extension of the normal field γ. Of course, v depends on the parameter ε, but an appropriate choice of f and g yields $|u - v| \leqslant C\varepsilon$, which leads to a gradient bound for u. Equivalently (as in [3, Sect. 1]), the maximum principle can be applied directly to the function

$$F(u)[(1 + |Du|^2)^{1/2} - Du \cdot \gamma \cos \theta]$$

where $d(x) = \operatorname{dist}(x, \partial\Omega)$ and F is an appropriate function which can be taken equal to 1 if H_z is bounded away from zero. The second approach, due to the present author, follows that in [6, Sect. 4] for uniformly elliptic equations and applies to problem (2) under appropriate hypotheses on a^{ij}, a, and b, which include problem (1)'. Now the maximum principle is applied to the function

$$w = |Du|^2 - |Du \cdot \gamma|^2 + f(d);$$

if H_z is not bounded away from zero, a preliminary change of dependent variable, equivalent to multiplication by $F(u)$ in Korevaar's method, must be performed. Since it can be shown that

$$\sup_{\Omega} |\gamma \cdot Du| \leqslant g_0 \sup_{\Omega} |Du| + g_1$$

for positive constants g_0 and g_1, with g_0 strictly less than one, the estimate on w leads to an estimate on $|Du|$. One technical detail to be noted is that this estimate is needed to prove the estimate on w. Both of these methods assume that Ω has a C^3 boundary, although $C^{2,\alpha}$ suffices; Ural'tseva's test function argument only needs C^2.

When $\partial\Omega$ has singularities, the maximum principle can sometimes be used to prove regularity of solutions of (1) or (1)'. In [4], Korevaar has obtained gradient bounds in this situation provided certain additional restrictions are placed on θ. These restrictions are given in terms of the geometry of the boundary at the point in question and known to be necessary for smooth solutions to exist.

So far, the behavior we have discussed can be verified in any number of space dimensions via the maximum principle. Higher regularity seems to be more delicate. If the data of the problem are sufficiently smooth, standard test function arguments [5, Chapter X] imply Hölder estimates for Du, and then Schauder theory gives further regularity.

In two dimensions, the author [7] has used maximum principle arguments (in conjunction with other techniques) to study the continuity of Du for problem (2). Under weak hypotheses on the coefficients, which reduce to Hölder continuity of H, θ, and γ in (1)', the function $g(x) = b(x, u, Du)$ can be extended in a natural way so that it satisfies a uniformly elliptic equation. It can then be shown that $g \leqslant Cd^\beta$ for some positive β. The precise form of the interior second derivative estimate for this problem gives a Hölder estimate for Du. The two-dimensional nature of this problem is used in several apparently crucial places. The same result for parametric surfaces was proved by Taylor [11], using geometric measure theory.

At a two-dimensional corner, Simon [10] used geometric measure theory to prove that the solution surface for (1)' has Hölder continuous normal provided the angle θ satisfies a certain condition at the corner. This condition guarantees that Du at the corner, if it exists, satisfies the two equations defined by taking the limit of the boundary conditions at the corner. Using Korevaar's gradient bound and the boundary Hölder gradient estimate mentioned above, this regularity of the solution surface can be proved via the maximum principle as well [9]. Suppose, for simplicity, that the corner is the point $(0, 0)$ and the two smooth portions of the boundary are Σ_1 lying on the x^2-axis and Σ_2 lying on the x^1-axis. (Note that the right angle can always be achieved by a linear change of independent variable.) Suppose also that the boundary condition can be written as $b^i(Du) = 0$ on Σ_i for $i = 1, 2$ and that $\det(\partial b^i/\partial p_j)$ is never zero (as is clear for the capillary problem). Then the functions $v_i(x) = b^i(Du)$ satisfy mixed boundary conditions, and, as in the previous paragraph, these functions also satisfy uniformly elliptic equations. Next, it follows from a simple barrier argument that the v's go to zero like a power of distance to the corner. Assuming that the equations $b^1(Du) = 0$ and $b^2(Du) = 0$ can be solved for Du (this is the condition on the contact angle previously referred to), the condition on the determinant provides the Hölder continuity of Du at the corner and the Hölder continuity nearby has already been shown.

References

1. P. Concus and R. Finn, *On capillary free surfaces in a gravitational field*, Acta Math. **132** (1975), 207–223.
2. N. Korevaar, *The normal variations technique for studying the shape of capillary surfaces*, Asterique **118** (1984), 189–195.
3. N. Korevaar, *An easy proof of the interior gradient bound for solutions to the prescribed mean curvature equation*, Proc. Symp. Pure Math. **45**—Part 2 (1986), 81–89.
4. N. Korevaar, *Maximum principle gradient estimates for the capillary problem* (to appear).
5. O.A. Ladyzhenskaya and N.N. Ural'tseva, *Linear and Quasilinear Elliptic Equations*, Academic Press, New York (1968).
6. G.M. Lieberman, *The nonlinear oblique derivative problem for quasilinear elliptic equations*, Nonlin. Anal. **8** (1984), 49–65.
7. G.M. Lieberman, *Two-dimensional nonlinear boundary value problems for elliptic equations*, Trans. Amer. Math. Soc. (to appear).
8. G.M. Lieberman, *Gradient estimates for capillary type problems via the maximum principle* (to appear).
9. G.M. Lieberman (to appear).
10. L.M. Simon, *Regularity of capillary surfaces over domains with corners*, Pac. J. Math. **88** (1980), 363–377.
11. J. Taylor, *Boundary regularity for solutions to various capillarity and free boundary problems*, Comm. Partial Differential Equations **2** (1977), 323–357.
12. N.N. Ural'ceva, *Solvability of the capillary problem. I, II*, Vestnik Leningrad Univ. Math. **6** (1979), 363–375; **8** (1980), 151–158.

Convex Functions Methods in the Dirichlet Problem for Euler– Lagrange Equations

Ilya J. Bakelman

In this paper we investigate *a priori* estimates for solutions of the second order elliptic E–L equations,* whose gradients satisfy some prescribed limitations. Such problems arise from the relativity theory and continuous mechanics and can be described in terms of variational problems for the n-dimensional multiple integrals

$$\int_B F(x, u, Du)\, dx, \tag{1}$$

whose integrands $F(x, u, p)$ are defined only for vectors p belonging to pre-scribed domain G in R^n. If G coincides with the whole space R^n, then we do not have any prescribed limitations for the gradient of desired solutions for the E–L equation corresponding to the functional (1). This most simple case was investigated in our paper [1].

We also consider applications of our estimates, derived in the present paper, to the Dirichlet problem for mean curvature equation in Euclidean and Minkowski spaces and for the equation of torsion of hardening rods. Finally, we review existence theorems of strong solutions in Hölder spaces for these problems related to differential geometry, physics, and mechanics.

1. The Main Geometric Constructions and Assumptions

1.1. The Main Notations

Let E^n, P^n, and Q^n be three n-dimensional Euclidean spaces with corre-sponding systems of Cartesian coordinates $x_1, x_2, \ldots, x_n$; $p_1, p_2, \ldots, p_n$; and $q_1, q_2, \ldots, q_n$. We will use the notations $x = (x_1, x_2, \ldots, x_n)$ for points of E^n;

---

*We will use the brief notation E–L equations for Euler–Lagrange equations.

$p = (p_1, p_2, \ldots, p_n)$ for points of P^n; and $q = (q_1, q_2, \ldots, q_n)$ for points of Q^n; and $(x, z) = (x_1, x_2, \ldots, x_n, z)$ for points of $E^n \times R$; $(p, w) = (p_1, p_2, \ldots, p_n, w)$ for points of $P^n \times R$.

Let $0', 0'', 0'''$ be origins in E^n, P^n, Q^n. We denote by $|e'|_{E^n}, |e''|_{P^n}$, and $|e'''|_{Q^n}$ the Lebesgue measures in E^n, P^n, and Q^n for the sets e', e'', e'''.

Let B be a bounded domain in E^n and let ∂B be a closed hypersurface in E^n. We denote by $\chi_z: B \to P^n$ the tangential mapping, generated by every function $z(x) \in C^1(B)$, i.e

$$p_i = \frac{\partial z}{\partial x_i}, \qquad i = 1, 2, \ldots, n. \tag{2}$$

Let G be a closed n-domain in P^n and the origin $0''$ is the interior point of ∂G. There are two possibilities, $\partial G \neq \varnothing$ and $\partial G = \varnothing$. In the first case, G can be either bounded or unbounded closed domain in P^n which is different from P^n, but in the second case $G = P^n$.

We denote by $U(\rho)$ the closed n-ball in P^n with the center $0''$ and the radius ρ. Let $r(G) = \sup \rho$, where sup is taken under the condition $U(\rho) \subset \operatorname{int} G$. If $\partial G \neq \varnothing$, then $r(G)$ is a positive finite number and

$$U(r(G)) \subset G. \tag{3}$$

If $G = \varnothing$, i.e. $G = P^n$, then $r(G) = +\infty$.

In many cases it is sufficient to consider only convex closed n-domains G. If G is the ball $|p| \leqslant a$, then $r(G) = a$ and $U(r(G)) = G$. In some problems it will be convenient to consider our problems in open domains int G.

Now let $\phi(p)$ be a convex (concave) function defined in the set int G. The cases when $\phi(p)$ can be extended to G or to a larger set $G' \supset G$ (for example to P^n) are not excluded. In this paper we assume that $\phi(p) \in C^2(\operatorname{int} G)$. We introduce the set function

$$\omega(\phi, e'') = |\chi_\phi(e'')|_{Q^n}, \tag{4}$$

where $\chi_\phi: \operatorname{int} G \to Q^n$ is the tangential mapping generated by the convex function $\phi(p)$ and e'' is any measurable subset of int G. The number $\omega(\phi, \operatorname{int} G)$ is called the *total area of the tangential mapping* generated by $\phi(p)$.

1.2. Estimators

Now we introduce the function $c(\rho)$ for $0 \leqslant \rho < r(G)$ by means of the formula

$$c(\rho) = \omega(\phi, U(\rho)). \tag{5}$$

Let

$$C(\phi) = \omega(\phi, U(r(G))) = \lim_{\rho \to r(G)} \omega(\phi, U(\rho)). \tag{6}$$

Since $\chi_\phi(U(r(G)))$ can be the whole space Q^n, then the case $C(\phi) = +\infty$ is not excluded. The function $c(\rho)$ is non-negative, continuous, and strictly increasing in $[0, r(G))$. Therefore there exists its inverse $\rho = b(t)$ for $0 \leqslant t < C(\phi)$.

1.3. Convex and Concave Supports of a Real Function

Let B be a bounded domain in E^n and let ∂B be a closed continuous hyper-surface in E^n. Let $u(x) \in C^2(B) \cap C(\bar{B})$. We denote by S_m the part of the graph of $u(x)$, located under the hyperplane $\gamma_m: z = m$ in $E^{n+1} = E^n \times R$, where $m = \inf u(x)$ and $M = \sup u(x)$ for $x \in \partial B$. Thus S_m is the subset of E^{n+1} consisting of all points $(x, u(x))$, where $x \in \bar{B}$ and $u(x) \leqslant m$. We introduce the following notations: $\bar{B}_m$ is the set in γ_m, whose projection on E^n coincides with $\bar{B}$; Γ_m and Γ are correspondingly the boundaries of H_m and H, where H_m and H are closed convex hulls of $\bar{B}_m$ and $\bar{B}$.

Since $u(x) \in C(\bar{B})$, then only the case

$$u_0 = \inf_{\bar{B}} u(x) < m \tag{7}$$

is interesting. Clearly $u_0 = u(x_0)$, where x_0 is an interior point of B. Let C_m be the closed convex hull of the set $\bar{B}_m \cup S_m$. Then

$$C_m = H_m \cup S_{v_1},$$

where S_{v_1} is the graph of a convex function $v_1(x) \in C(H)$ such that

$$v_1(x) = m \tag{8}$$

for all $x \in \partial H$ and

$$v_1(x) \leqslant u(x) \tag{9}$$

for all $x \in \bar{B}$. The convex function $v_1(x)$ is called the *convex support* of the function $u(x)$.

If S_M is the part of the graph of the function $u(x)$, located over the hyperplane $\gamma_M: z = M$, then the analogical geometric constructions lead to the *concave support* $v_2(x) \in C(H)$ of the function $u(x)$. Clearly $v_2(x) = M$ for all $x \in \partial H$ and $v_2(x) \lrcorner \geqslant u(x)$ for all $x \in \bar{B}$.

For general convex and concave functions there is the concept of the normal mapping, which extends the concept of the tangential mapping in terms of supporting hyperplanes to the graphs of convex or concave functions. We will use the notation $\chi_v: H \to P^n$ for the normal mapping of convex or concave functions $v(x) \in C(H)$ (for definition and properties, see [1], [2]).

1.4. Two-Sided Estimates for the Functions $u(x) \in C^2(B) \cap C(\bar{B})$

Our first theorem is related to the properties of arbitrary functions $u(x) \in C^2(B) \cap C(\bar{B})$ satisfying the limitation

$$\chi_u(B) \subset G. \tag{10}$$

We recall that (10) is a real limitation if $\partial G \neq \varnothing$. For every such function $u(x)$ the compositions $\chi_\phi \circ \chi_{v_i}: \text{int } H \to Q^n$, $i = 1, 2$, can be constructed.

Clearly B is the open subset of the convex set $\text{int } H$. We introduce the numbers

$$\omega_+ = \omega(\phi, \chi_{v_1}(B)) = |(\chi_\phi \circ \chi_{v_1})(B)|_{Q^n}, \tag{11}$$

$$\omega_- = \omega(\phi, \chi_{v_2}(B)) = |(\chi_\phi \circ \chi_{v_2})(B)|_{Q^n}, \tag{12}$$

Then the following Theorem takes place (see the proof in [7]).

Theorem 1. *Let* $u(x) \in C^2(B) \cap C(\bar{B})$ *and let the inclusion* (10) *be fulfilled. If the inequalities*

$$\omega_+ < C(\phi), \tag{13}$$

$$\omega_- < C(\phi), \tag{14}$$

hold, then

$$m - b(\omega_+)\text{diam } B \leqslant u(x) \leqslant M + b(\omega_-)\text{diam } B \tag{15}$$

for all $x \in \bar{B}$. *If* $\partial G \neq \varnothing$, *then inequalities* (13) *and* (14) *can be extended to* $\omega_\pm \leqslant C(\phi)$.

1.5. Assumptions with Respect to Euler–Lagrange Equations

Let $F(x, u, p)$ be a C^2-function in $\bar{B} \times R \times G$. It is well known that any C^2-function, minimizing (maximizing) the multiple integral (1), satisfies the following E–L equation:

$$\sum_{i=1}^{n} \frac{d}{dx_i}\left(\frac{\partial F}{\partial p_i}\right) - \frac{\partial F}{\partial u} = 0. \tag{16}$$

We rewrite equation (16) in the form

$$\sum_{i,k=1}^{n} \frac{\partial^2 F(x, u, Du)}{\partial p_i \partial p_k} u_{ik} = D(x, u, Du), \tag{17}$$

where

$$D(x, u, p) = \frac{\partial F}{\partial u} - \sum_{i=1}^{n} \frac{\partial^2 F}{\partial p_i \partial u} p_i - \sum_{i=1}^{n} \frac{\partial^2 F}{\partial p_i \partial x_i}. \tag{18}$$

Equation (17) is elliptic if and only if the quadratic form

$$\sum_{i,k=1}^{n} \frac{\partial^2 F}{\partial p_i \partial p_k} \xi_i \xi_k \tag{19}$$

is positive (negative) definite in $\bar{B} \times R \times G$. If G is different from P^n, we have the *variational problem with limitation on the gradient of solutions.*

Now let $F(x, u, p)$ be a C^2-function in $\bar{B} \times R \times G$ and let the quadratic form (1.8) be positive definite in $\bar{B} \times R \times G$. Then there exist functions $g_1(x, u)$, $g_2(x, u)$, $R_1(p)$, $R_2(p)$, $l_1(x, u)$, $l_2(x, u)$, $\phi_1(p)$, $\phi_2(p)$ such that the following conditions hold:

A.1. The functions $R_1(p), R_2(p)$ are positive and locally summable in G with degree n; the functions $g_1(x, u), g_2(x, u)$ are non-negative in $B \times R$ and the inequalities

$$D(x, u, p) \leqslant \frac{g_1(x, u)}{R_1(p)} \tag{20}$$

and

$$-D(x, u, p) \leqslant \frac{g_2(x, u)}{R_2(p)} \tag{21}$$

correspondingly hold in every point $(x, u, p) \in B \times R \times G$ if $D(x, u, p) \geqslant 0$ or $D(x, u, p) < 0$.

A.2. $\phi_1(p)$, $\phi_2(p)$ are strictly convex C^2-functions in G; $l_1(x, u)$, $l_2(x, u)$ are positive in $\bar{B} \times R$, and the inequalities

$$l_k(x, u) \det\left(\frac{\partial^2 \phi_k(p)}{\partial p_i \partial p_j}\right) \leqslant R_k^n(p) \det\left(\frac{\partial^2 F(x, u, p)}{\partial p_i \partial p_j}\right), \tag{22}$$

$k = 1, 2$, hold in $\bar{B} \times R \times G$.

A.3. The functions

$$\psi_k(x, u) = \left[\frac{g_k(x, u)}{l_k(x, u)}\right]^n, \tag{23}$$

$k = 1, 2$, are summable in B for every fixed $u \in R$ and are non-decreasing with respect to u for every fixed $x \in \bar{B}$.

Remark. The inequalities (22) will have more general and traditional form if we replace $\det(\phi_{ij}(p))$ by some positive function $T(p)$. But existence theorems for generalized and smooth solution of Monge–Ampere equations

$$\det(\phi_{ij}(p)) = T(p)$$

(see [3], [4], [5], [6]) provide without losing generality to use Hessians of convex functions $\phi(p)$ in inequalities (22).

EXAMPLE. Let

$$F(x, u, p) = \phi(p) + f(x, u)$$

for all $(x, u, p) \in \bar{B} \times R \times G$ and $\phi(p) \in C^2(\text{int } G)$. Then equation (16) becomes

$$\sum_{i,k=1}^n \frac{\partial^2 \phi(Du)}{\partial p_i \partial p_k} u_{ik} = f_u(x, u). \tag{24}$$

Let

$$f_u(x, u) = f_u^+(x, u) - f_u^-(x, u), \tag{25}$$

where $f_u^+(x, u)$ and $f_u^-(x, u)$ are correspondingly the positive and negative parts of $f_u(x, u)$. Thus

$$g_1(x, u) = f_u^+(x, u), \qquad g_2(x, u) = f_u^-(x, u) \tag{26}$$

and

$$R_1(p) = R_2(p) = 1. \tag{27}$$

Clearly

$$\phi_1(p) = \phi_2(p) = \phi(p) \tag{28}$$

and

$$l_1(x, u) = l_2(x, u) = 1. \tag{29}$$

From (26–29) it follows that only Assumption A.3 has non-trivial statement, which can be formulated in the following way. The functions $f_u^+(x, u)$ and $f_u^-(x, u)$ belong to $L^n(B)$ for every fixed $u \in R$ and these functions are non-decreasing with respect to u for every fixed $x \in \bar{B}$.

Important applications to problems of differential geometry and mechanics will be considered below.

2. Two-Sided Estimates of Solutions of the Dirichlet Problem for Elliptic Euler–Lagrange Equations

2.1. The Main Theorem on Two-Sided Estimates

Theorem 2. *Let* $u(x) \in C^2(B) \cap C(\bar{B})$ *be a solution of E–L equation* (16), *satisfying the condition*

$$\chi_u(B) \subset G \text{ and } \partial G \neq \varnothing$$

We assume that Assumptions A.1–A.3 (see subsection 1.5) are fulfilled and the inequalities

$$\Omega_+ \leqslant C(\phi_1), \tag{30}$$

$$\Omega_- \leqslant C(\phi_2) \tag{31}$$

hold, where

$$\Omega_+ = \frac{1}{n^n} \int_B \psi_1(x, m) \, dx, \qquad \Omega_- = \frac{1}{n^n} \int_B \psi_2(x, M) \, dx; \tag{32}$$

$$m = \inf_{\partial B} u(x), \qquad M = \sup_{\partial B} u(x), \tag{33}$$

and $C(\phi_k)$, $k = 1, 2$, *are defined by* (6).
   Then the inequalities

$$m - b_1(\Omega_+)\text{diam } B \leqslant u(x) \leqslant M + b_2(\Omega_-)\text{diam } B \tag{34}$$

hold for all $x \in \bar{B}$. *If* $\partial G = \varnothing$, *then* (30), (31) *have to be replaced by* $\Omega_+ < C(\phi_1)$, $\Omega_- < C(\phi_2)$.

Comments to Theorem 2. (a) According to the definition of the numbers ω_+ and ω_- (see subsection 1.4) we establish the inequalities

$$\omega_+ \leqslant \Omega_+, \qquad \omega_- \leqslant \Omega_-. \tag{35}$$

Since $b_1(t)$, $b_2(t)$ are strictly increasing functions, the direct application of Theorem 1 completes the proof of Theorem 2.

The proof of inequalities (35) is based on the analytic expression for the numbers ω_+ and ω_- and the fact that $u(x)$ is the solution of equation (16), satisfying the condition

$$\chi_u(B) \subset G.$$

For details see [7].

(b) The particular case of Theorem 2, where $G = P^n$ (i.e. $\partial G = \varnothing$), was considered in [1]. If $\partial G \neq \varnothing$ then essentially new considerations are required.

2.2. Total Area of the Tangential Mapping of a Complete Infinite Convex Hypersurface $w = \phi(p)$

Here we assume that G is either the n-ball; $|p| \leqslant a, (0 < a < +\infty)$, or the entire space P^n. Then

$$U(r(G)) = G. \tag{36}$$

Hence

$$C(\phi) = \omega(\phi, \text{int } G), \tag{37}$$

where $\omega(\phi, \text{int } G)$ is the total area of the tangential mapping of a convex hypersurface $w = \phi(p)$.

Let $w = \phi(p)$ be a complete convex hypersurface and K_ϕ be the asymptotic cone of ϕ (see details in [1]). Since $\chi_\phi(G) = \chi_{K_\phi}(G)$, then

$$\omega(\phi, \text{int } G) = |\chi_{K_\phi}(G)|_{Q^n}. \tag{38}$$

If G is a bounded set, then K_ϕ is necessarily a ray orthogonal to E^n and $\chi_{K_\phi}(G) = Q^n$. If K_ϕ is projected one-to-one on E^n, then K_ϕ is a non-degenerate n-dimensional convex cone and $\chi_{K_\phi}(G)$ is a bounded closed set in Q^n. Hence the origin $0''' \in \text{int } \chi_{K_\phi}(G)$ and $|\chi_{K_\phi}(G)|_{Q^n} < +\infty$.

2.3. Hypersurfaces with Prescribed Mean Curvature in Euclidean Space E^{n+1}

Let the hypersurface S with prescribed mean curvature $H(x)$ be a graph of the function $u(x) \in C^2(B) \cap C(\overline{B})$. If $H(x) \in C(\overline{B})$, then $u(x)$ satisfies the E–L equation for the functional

$$I(u) = \int_B [\sqrt{1 + (Du(x))^2} + nH(x)u(x)]\, dx.$$

According to the example considered in subsection 1.5, we obtain $\phi_1(p) = \phi_2(p) = \sqrt{1 + |p|^2}$; $g_1(x, u) = H_+(x)$, $g_2(x, u) = H_-(x)$; $l_1(x, u) = l_2(x, u) = 1$:

$R_1(p) = R_2(p) = 1$ for all $x \in \bar{B}$, $u \in R$, and $p \in P^n$, where $H_+(x)$ and $H_-(x)$ are correspondingly positive and negative parts of the function $H(x)$.

The equation of the convex cone K_ϕ is $w = |p|$ for all $p \in P^n$. Hence $\chi_{K_\phi}(P^n)$ is the unit n-ball $|q| \leqslant 1$ in Q^n. If μ_n is the volume of this ball, then inequalities (30) and (31) become

$$\Omega_\pm = \int_B H_\pm^n(x)\,dx < \mu_n \tag{39}$$

because $\omega(\phi, P^n) = \mu_n$. The elementary calculation shows that

$$b_1(t) = b_2(t) = [1 - (t/\mu_n)^{2/n}]^{1/2}$$

for all $0 \leqslant t < \mu_n$, where $b_1(t)$ and $b_2(t)$ are estimators in inequalities (34).

Let B be the n-dimensional ball of the radius r_0 and let $H(x) = H_0 = \text{const} > 0$. Since the Dirichlet problem for mean curvature equation has not more than one solution for functions $u(x) \in C^2(B) \cap C(\bar{B})$, then the conditions

$$r_0 H_0 < 1 \quad \text{and} \quad 0 < \mu_n \tag{40}$$

are necessary and sufficient for solvability of the Dirichlet problem for constant mean curvature equation by zero boundary data in the space $C^2(\bar{B})$; moreover the suitable spherical segment is a unique solution of this problem. Inequalities (30)–(31) become (40) for the constant mean curvature equation.

2.4. Spacelike Hypersurfaces with Prescribed Mean Curvature in Minkowski Space M^{n+1}

The space $R^{n+1} = \{(x, t)\} = \{(x_1, x_2, \ldots, x_n, z)\}$ with the metric

$$ds^2 = \sum_{i=1}^n (dx_i)^2 - dz^2 \tag{41}$$

is called the Minkowski space and is denoted by M^{n+1}. Let S be a hypersurface such that ds^2 restricts to a positive definite form on S. Such S is called spacelike. If S is the graph of a function $z = u(x)$, then S is spacelike if and only if $|Du(x)| < 1$ for any x belonging to the domain of the function $u(x)$. The spacelike hypersurfaces were studied by Calabi [8] and Cheng and Yau [9] in connection with Bernstein conjecture in M^{n+1}. The spacelike solutions $u(x) \in C^2(\bar{B})$ of E–L equation for the functional

$$I_1(u) = \int_B [(1 - (Du)^2)^{1/2} - nH(x)u]\,dx$$

have prescribed mean curvature $H(x)$ in M^{n+1}. Clearly $H(x) \in C(\bar{B})$ for such hypersurfaces. According to Assumptions A.1–A.3 we have

$$\phi_1(p) = \phi_2(p) = (1 - |p|^2)^{1/2}$$

for all $p \in \text{int } G$, where int G is the n-unit ball $|p| < 1$ in P^n. It is also clear that

$g_1(x, u) = H_+(x)$, $g_2(x, u) = H_-(x, u)$; $l_1(x, u) = l_2(x, u) = 1$; $R_1(p) = R_2(p) = 1$ for all $x \in \bar{B}$, $u \in R$, and $P \in \operatorname{int} G$. Clearly $\chi_\phi(\operatorname{int} G) = Q^n$. Therefore $H_\pm(x) \in L^n_{loc}(B)$ and inequalities (30)–(31) become

$$\int_B H_\pm^n(x)\, dx \leqslant +\infty.$$

Thus there is an essential difference between the solutions of the Dirichlet problem for the mean curvature equation in Euclidean and Minkowski spaces.

3. Existence Theorem for Torsion of Hardening Rods

Let P be a prismatic rod represented by the cylinder with the base $\bar{B} = B \cup \partial B$ and generators parallel to the z-axis, where B is a bounded domain in the x, y-plane. Let the base of P be clamped and let the rod P twist under the action of a moment M. We denote by w the torsion per unit length of the rod. Let $u(x, y)$ be the stress function of the rod P, then $T^2 = (u_x^2 + u_y^2)^{1/2}$ is the intensity of the tangential stress and

$$\frac{\partial}{\partial x}\left(g(T^2)\frac{\partial u}{\partial x}\right) + \frac{\partial}{\partial y}\left(g(T^2)\frac{\partial u}{\partial y}\right) = -2w \tag{42}$$

is the equation of hardening rods.

The function $g(T^2)$ is called the modulus of plasticity of the rod P. It describes the dependence between the intensity of the shear strain tensor Γ and the intensity of the tangential stress T. The experimental law $d\Gamma/dT > 0$ is the necessary and sufficient condition of ellipticity for equation (42). The problem of the torsion of hardening rods can be reduced to the Dirichlet problem for equation (42) with zero boundary data.

Below we formulate one of the series of existence theorems proved in [10] for equations which are slightly generalized from (42) to n dimensions.

Theorem 3. *Consider the Dirichlet problem*

$$\sum_{i=1}^{n} \frac{\partial}{\partial x_i}\left(g(T^2)\frac{\partial u}{\partial x_i}\right) = nH(x), \tag{43}$$

$$u = 0 \qquad \text{for all } x \in \partial B, \tag{44}$$

where the following conditions are fulfilled:

(1) *The domain B in E^n is bounded and simply connected, and ∂B is a $C^{m+2,a}$ closed hypersurface ($m \geqslant 1, 0 < a < 1$).*

(2) *$g(T^2) > 0$ for every T and $g(T^2)$ as a function of the variable $\xi = T^2 \in [0, +\infty)$ has $(m+1)$ continuous derivatives on $[0, +\infty)$ and its $(m+1)$-th derivative satisfies the Hölder condition of degree a on $[0, +\infty)$, where $m \geqslant 1$ and $0 < a < 1$.*

(3) *The following inequalities*

$$g(T^2) \geqslant C_0(1 + T^2)^{\gamma/2},$$

$$C_1 g(T^2) \geqslant \frac{d\Gamma}{dT} \geqslant C_2(1 + T^2)^{\gamma-2/2} \left| \frac{d^2\Gamma}{(dT)^2} \right| \leqslant C_3(1 + T^2)^{\gamma-3/2}$$

hold for all $T \in [0, +\infty)$, *where* $T^2 = u_{x_1}^2 + \cdots + u_{x_n}^2$, $\Gamma = g(T^2)T$, $\gamma = \text{const} \geqslant -1$; C_0, C_1, C_2 *are positive constants, and* C_3 *is a non-negative constant.*

(4) $$\int_B H_\pm^n(x)\, dx < n\mu_n \Gamma^n(+\infty). \qquad (45)$$

(*Note that the general inequalities* (30)–(31) *turn into inequalities* (45).)

(5) $$S_{n-1}(x, B) \geqslant \begin{cases} 0 & \text{if } \gamma = \text{const} > -1, \\[2mm] \dfrac{nh(x)}{(n-1)C_0} & \text{if } \gamma = -1, \end{cases}$$

where $h(x) = \max(H_+(x), H_-(x))$, $S_{n-1}(x, \partial B)$ *is the mean curvature of* ∂B *at the point* $x \in \partial B$ *in the interior normal direction.*

Then the Dirichlet problem (43)–(44) has a unique solution $u(x) \in C^{m+2,a'}(\bar{B})$, where $0 < a' \leqslant a < 1$.

Note that the existence Theorem for mean curvature equation in E^{n+1} is a particular case of Theorem 3.

Acknowledgments. Research was supported by NSF Grant 8420850.

References

1. I. Bakelman, *The boundary value problems for non-linear elliptic equations and the maximum principle for Euler–Lagrange equations*, Arch. Ration. Mechanics & Analysis, **93**, No. 3, (1986), 271–300.
2. I. Bakelman, *R-curvature, estimates and stability of solutions of the Dirichlet problem for elliptic equations*, Journ. of Diff. Equations, **43**, No. 1, (1982), 106–133.
3. A. Pogorelov, *The Minkowski multidimensional problem*, New York, J. Wiley (1978).
4. S.Y. Cheng and S.T. Yau, *On the regularity of the solution of the Monge–Ampere equations* $\det(u_{ij}) = F(x, u)$, Comm. Pure and Appl. Mathem. **29** (1976), 495–216.
5. I.J. Bakelman, *Generalized elliptic solutions of the Dirichlet problem for n-dimensional Monge–Ampere equations*, Proc. of Symposia in Pure Mathem. **44** (1985), 1–30.
6. L. Caffarelli, L. Nirenberg, and J. Spruck, *The Dirichlet problem for non-linear Second Order Elliptic Equations 1. Monge–Ampere equations*, Comm. Pure and Appl. Mathem. **37** (1984), 369–402.
7. I. Bakelman, *Estimates of solutions for elliptic Euler–Lagrange equations and related topics in diff. geometry of convex functions*, IMA Preprint Series **209** (1986), 1–23.

8. E. Calabi, *Examples of Bernstein problems for some nonlinear equations*, Proc. Symp. Global Analysis, University of California, Berkeley (1968).

9. S.Y. Cheng and S.T. Yau, *Maximal spacelike hypersurfaces in the Lorentz–Minkowski spaces*, Annals of Mathem. **104** (1976), 407–419.

10. I. Bakelman, *Notes concerning torsion of hardening rods and its generalizations*, IMA Preprint Series **208** (1986), 1–38.

Stability of a Drop Trapped Between Two Parallel Planes: Preliminary Report

Thomas I. Vogel

1. Introduction

This paper deals with the physical problem of a drop of liquid trapped between two homogeneous parallel planes in the absence of gravity. Explicit stability results are derived in the case of the contact angles between the liquid and the planes being equal to $\pi/2$, and a sufficient condition for stability for more general contact angles is derived. The present paper is an abbreviated version of a paper which will appear [7]. The abbreviation consisted of omitting most proofs. I have been informed by Professor Stephan Hildebrandt that his student, Maria Athanassenas, has independently derived the stability results for contact angles equal to $\pi/2$, apparently using different methods.

The problem of the trapped drop is a capillary problem, since it involves the interaction of surface tension and wetting energy. It is related to symmetric unbounded liquid bridges [8], and pendent drops ([1], [9], and [10], for example). The factor complicating the trapped drop problem is the presence of two wetted areas and two contact angles. This makes even the case without gravity and with contact angles $\pi/2$ interesting.

The energy functional associated with the problem of a drop trapped between two parallel planes Π_1 and Π_2 in the absence of gravity is $E(\Omega) = A(\Sigma) - a_1 A(S_1) - a_2 A(S_2)$, where Ω is the region in space occupied by the drop, Σ is the liquid-air interface, S_i is the region wetted by the drop on plane Π_i, a_1 and a_2 are constants depending on the materials involved, and A denotes area. Since the volume of the drop is constrained we seek local minima of $E(\Omega)$ in the family of sets Ω with volume$(\Omega) = V$, a constant.

The associated Euler–Lagrange equations yield that the mean curvature of the free surface Σ is a constant H, and that the angle between the normal to Σ and the normal to Π_i along their curve of contact is $\cos^{-1}(a_i)$. A standard symmetrization argument shows that Ω is rotationally symmetric (see, e.g. [10]).

Rotationally symmetric surfaces of constant mean curvature were studied by Delauney [2], who showed that they are obtained by tracing a focus of a conic section which is rolled without slipping along a line, and revolving the resulting curve around the line. In this paper we will be concerned with unduloids, which are generated by rolling ellipses. These have been used by Concus and Finn [1] in the study of pendent drops.

2. Formulation of the Variational Problem and the Stability Criterion

We wish to minimize

$$I(f) = 2\pi \int_0^h f\sqrt{1 + (f')^2}\, dx - a_1\pi[f(0)]^2 - a_2\pi[f(h)]^2$$

over the set of C^1 functions f which satisfy

$$J(f) = \pi \int_0^h f^2\, dx = V.$$

Pick $\eta_1, \eta_2 \in C^1([0, h])$. We need

$$I(\varepsilon_1, \varepsilon_2) = I(f + \varepsilon_1\eta_1 + \varepsilon_2\eta_2)$$

to have a local minimum at $(0, 0)$ subject to the constraint

$$J(\varepsilon_1, \varepsilon_2) = J(f + \varepsilon_1\eta_1 + \varepsilon_2\eta_2) = V$$

independently of η_1 and η_2.

Omitting details, the first variation yields the Euler–Lagrange equations:

$$H = \frac{1}{2}\left(\frac{f''}{(1 + (f')^2)^{3/2}} - \frac{1}{f(1 + (f')^2)^{1/2}}\right) \tag{2.1}$$

for some Lagrange multiplier H (the right hand side is the mean curvature operator), with boundary conditions

$$\frac{f'(0)}{\sqrt{1 + (f'(0))^2}} = a_1, \quad \text{and}$$

$$\frac{f'(h)}{\sqrt{1 + (f'(h))^2}} = -a_2. \tag{2.2}$$

The boundary conditions correspond to prescribed contact angles.

We now impose the sufficient criterion for the solution of a Lagrange multiplier problem to be a local minimum (see [3] or [6]). In the case we consider, this reduces, after some manipulation, to the following.

Theorem 2.1. *If $f(x)$ is a solution to (2.1) and (2.2), if $\pi \int_0^h f^2 \, dx = V$, and if the quadratic form*

$$\beta(\psi) = \int_0^h \frac{f(\psi')^2}{(1 + (f')^2)^{3/2}} - \frac{\psi^2}{f(1 + (f')^2)^{1/2}} \, dx$$

is positive definite on the space of functions

$$f^\perp = \left\{ \psi \in C^1([0, h]) \,\middle|\, \int_0^h \psi f \, dx = 0 \right\},$$

then f is a local minimum of the energy functional $I(f)$ subject to the constraint $J(f) = V$. If $\beta(\psi) < 0$ for any $\psi \in f^\perp$, then f is not a local minimum.

3. Analysis of the Stability Criterion

We now consider quadratic forms of the type

$$\beta(\psi) = \int_0^h (P(\psi')^2 + Q\psi^2) \, dx$$

on subspaces of $C^1([0, h])$ of the form $f^\perp = \{\psi \mid \int_0^h \psi f \, dx = 0\}$, where f is a given function. Here P is a strictly positive function of x, and Q is a strictly negative function of x. Consider the Sturm–Liouville eigenvalue problem

$$L(\psi) = -(P\psi')' + Q\psi = \lambda\psi, \qquad \text{with } \psi'(0) = \psi'(h) = 0. \tag{3.1}$$

Since this is a regular Sturm–Liouville system, the sequence $\{\lambda_i\}, i = 0, 1, \ldots$ of eigenvalues is strictly increasing. Since $Q(x) < 0$ for all x, it follows that there is at least one negative eigenvalue, since β, the associated quadratic form, is negative for $\psi \equiv 1$. If we have $\lambda_1 \leqslant 0$, $\beta(\psi)$ is obviously not positive definite on $f^\perp$, since $f^\perp$ has codimension 1 and the space spanned by the first two eigenfunctions ψ_0 and ψ_1 has dimension 2. The interesting case is when $\lambda_0 < 0 < \lambda_1$. The following theorem can be proven using linear algebra.

Theorem 3.1. *Suppose $\lambda_0 < 0 < \lambda_1$, and let $\phi(x)$ be the unique function which solves $L(\phi) = f$ with $\phi'(0) = \phi'(h) = 0$. (That such a ϕ exists is shown in [5].) Then the following statements are equivalent:*

1. $\int_0^h \phi f \, dx < 0$;
2. $\beta(\psi)$ *is positive definite on $f^\perp$.*

Note 3.2. We can also consider these other boundary conditions in (3.1):

$$\psi(0) = \psi'(h) = 0, \quad \text{or}$$
$$\psi'(0) = \psi(h) = 0, \quad \text{or}$$
$$\psi(0) = \psi(h) = 0.$$

Theorem 3.1 will still hold, except that the boundary conditions on ϕ will change correspondingly. We will not need this in this paper.

4. Stability of the Trapped Drop for $a_1 = a_2 = 0$

We will now apply Theorem 3.1 to the problem of the trapped drop. First, however, we must find a more convenient characterization for ϕ. Suppose that $f(x; \varepsilon)$ is a family of functions smoothly parameterized by ε which solve equation (2.1) with mean curvature $H(\varepsilon)$. Suppose also that $f'(0; \varepsilon)$ and $f'(h; \varepsilon)$ are constant in this family. Then one can verify that

$$L(f_\varepsilon) = -2f \frac{dH}{d\varepsilon},$$

where f_ε is $\partial f(x; \varepsilon)/\partial \varepsilon$, and L is as in Section 3, with

$$P = \frac{f}{(1 + (f')^2)^{3/2}} \quad \text{and} \quad Q = \frac{-1}{(1 + (f')^2)^{1/2}},$$

and that $f_\varepsilon'(0) = f_\varepsilon'(h) = 0$. It follows that the ϕ of Section 3 is

$$\frac{-1}{2\left(\dfrac{dH}{d\varepsilon}\right)} f_\varepsilon.$$

The condition that $\int_0^h \phi f \, dx < 0$ becomes the requirement that

$$\frac{dH}{d\varepsilon} \int_0^h f f_\varepsilon \, dx > 0.$$

If we set $V(\varepsilon) = \pi \int_0^h f^2(x; \varepsilon) \, dx$, we obtain the following.

Theorem 4.1. *Suppose $f(x)$ solves (2.1) and (2.2) with $J(f) = V$. Then f will be a local minimum for the energy functional $I(f)$ subject to the constraint $J(f) = V$ if:*

(a) *The eigenvalues of the Sturm–Liouville problem*

$$L(\psi) = -\left(\frac{f\psi'}{(1 + (f')^2)^{3/2}}\right)' - \frac{\psi}{f(1 + (f')^2)^{1/2}} = \lambda\psi, \qquad (4.1)$$

$$\psi'(0) = \psi'(h) = 0$$

satisfy $\lambda_0 < 0 < \lambda_1 < \dots$, and

(b) *$f(x) \equiv f(x; \varepsilon)$ may be embedded in a smoothly parameterized family $f(x; \varepsilon)$ of solutions of (2.1) and (2.2) with fixed values of a_1 and a_2 and*

$$\left.\frac{dH}{d\varepsilon}\right|_{\varepsilon=\varepsilon_0} \left.\frac{dV}{d\varepsilon}\right|_{\varepsilon=\varepsilon_0} > 0.$$

If $\lambda_1 < 0$, then $f(x)$ is not stable, and if $\lambda_0 < 0 < \lambda_1$ but

$$\frac{dH}{d\varepsilon}\bigg|_{\varepsilon=\varepsilon_0} \frac{dV}{d\varepsilon}\bigg|_{\varepsilon=\varepsilon_0} < 0,$$

then $f(x)$ is also unstable.

We will now apply this theorem to characterize completely the stable solutions in the case $a_1 = a_2 = 0$. The solutions to (2.1) and (2.2) may be classified into families F_k, $k = 0, 1, 2, \ldots$, where F_0 is the set of cylinders, and for $k \geq 1$, F_k is the set of unduloids generated by rolling ellipses of arc length $2h/k$ and varying eccentricity along the axis of symmetry. For each of these ellipses we obtain two members of F_k, one with a local maximum at $x = 0$ and the other with a local minimum there. As the eccentricity of the generating ellipse tends to unity, the members of F_k approach a configuration of spheres and half-spheres with a total volume of $(2\pi/3k^2)h^3$. The other limit, as eccentricity tends to zero, is a cylinder of volume $h^3/k^2\pi^2$. Thus the families F_k, $k = 1, 2, \ldots$ bifurcate from F_0.

Theorem 4.2. *Consider the Sturm–Liouville problem (4.1) with $f(x) \equiv c$. Then the number of non-positive eigenvalues is $[(h/\pi c) + 1]$, where $[x]$ is the greatest integer function. Moreover, if $c > h/\pi$, then $f(x) \equiv c$ is a local minimum for $I(f)$ subject to the constraint $J(f) = \pi c^2 h$.*

PROOF. This is a straightforward computation. □

Theorem 4.3. *All members of the families F_k, $k = 1, 2, \ldots$ are unstable.*

PROOF. A bifurcation of the solutions to (2.1) and (2.2) will occur if and only if 0 is an eigenvalue of L (see [4]). It is clear that the only bifurcations occur when F_k, $k = 1, 2, \ldots$, splits off from F_0; there are no secondary bifurcations. From our analysis of F_0, it follows that each element of the family F_k, $k = 1, 2, \ldots$, will yield an L with at least k negative eigenvalues, proving that F_k, $k = 2, 3, \ldots$ is unstable. A separate argument is needed for F_1.

An explicit calculation shows that members of F_1 generated by ellipses of very large eccentricity are unstable with respect to perturbations which wet the drier wall. Thus, either these unduloids lead to two negative eigenvalues in the Sturm–Liouville problem (4.1), or $(dH/d\varepsilon)(dV/d\varepsilon) \leq 0$ for these unduloids. However, one can also verify by an explicit construction that $(dH/d\varepsilon)(dV/d\varepsilon) > 0$ for unduloids of F_1 generated by ellipses of sufficiently large eccentricity. It therefore follows that for these unduloids (4.1) has two negative eigenvalues. The number of eigenvalues cannot change without a bifurcation occurring, thus the entire family F_1 is unstable. □

References

1. P. Concus and R. Finn, *The shape of a liquid pendent drop*, Phil. Trans. Roy. Soc. London **292**, no. 1391 (1979), 307–340.
2. C. Delaunay, *Sur la surface de révolution dont la courbure moyenne est constante*, J.

Math Pures & Appl. **6** (1841), 309–320.

3. R.P. Gillespie, *Partial Differentiation*, Oliver and Boyd, Ltd. (1951).

4. W.F. Langford, *Bifurcation Theory of Nonlinear Boundary Value Problems*, Thesis, California Institute of Technology (1971).

5. B.M. Leviton and I.S. Sargsjan, *Introduction to Spectral Theory*, Translations of Mathematical Monographs **39**, American Math Soc., Providence, RI (1975).

6. R.M. Meyer, *Essential Mathematics for Applied Fields*, Springer-Verlag, New York (1979).

7. T.I. Vogel, *Stability of a drop trapped between two parallel planes* (to appear).

8. T.I. Vogel, *Symmetric unbounded liquid bridges*, Pac. J. Math **103**, no. 1 (1982), 205–241.

9. H. Wente, *The stability of the axially symmetric pendent drop*, Pac. J. Math **88**, no. 2 (1980), 421–470.

10. H. Wente, *The symmetry of sessile and pendent drops*, Pac. J. Math **88**, no. 2 (1980), 387–397.

The Limit of Stability of Axisymmetric Rotating Drops

Frederic Brulois

1. Introduction

We consider an incompressible liquid drop Ω held together by surface tension in the absence of gravity; the drop is assumed to be small enough to make self-gravitation effects negligible. It is further assumed that the drop is *driven* to rotate at an imposed angular velocity. Finally, consideration is restricted to drop shapes Ω which are simply-connected and axisymmetric with respect to the axis of rotation.

The configurations Ω in gyrostatic equilibrium have been well studied. For background information, we refer the reader, for example, to the papers of Plateau [5], Rayleigh [7], Poincaré [6], Appell [1], Chandrasekhar [3], and Brown and Scriven [2]. Up to a normalization, these configurations form a one-parameter family; a convenient parameter is the dimensionless

$$C = \frac{(\text{density})(\text{angular velocity})^2(\text{equatorial radius})^3}{8\,(\text{surface tension})}, \tag{1}$$

which measures the "rotation rate" (and is denoted by Σ by Chandrasekhar [3]).

The stability of the equilibrium configurations has been studied numerically, notably by Chandrasekhar [3] and Brown and Scriven [2]. Chandrasekhar used his method of the virial, preserving only low order terms; he obtained, for the limit of stability C_0, the approximate value

$$C_0 \simeq .458706.$$

Brown and Scriven contributed to the subject a comprehensive numerical analysis—based on the finite element method—of the axisymmetric family as well as the bifurcating families and their stability to different types of perturbations; this for the present problem of a *driven* drop, as well as for the—physically more realistic—problem of an *isolated* drop, i.e. a drop for which the angular momentum is imposed instead of the angular velocity. They obtained the following approximate value:

$$C_0 \simeq .4449.$$

In this paper, a simple procedure to obtain arbitrarily precise estimates for C_0 will be presented; e.g. it is shown that

$$.448862813 < C_0 < .448862814. \tag{2}$$

However, there are still two gaps in this mathematical proof, as will be noted later, and consequently this paper is a preliminary outline of results to be published in more detail and more completely elsewhere.

2. The Variational Problem

We are led to the following variational problem:

(i) minimize the energy functional (actually the energy of the system divided by the surface tension):

$$\text{(area of } M) - 4C \text{ (moment of inertia of } \Omega),$$

where $M = \partial\Omega$ and C is the above Chandrasekhar number (1);
(ii) subject to the constraints:

$$\text{(volume of } \Omega) = \text{constant}, \tag{3}$$

$$\text{(centroid of } \Omega) = \text{origin}. \tag{4}$$

We now consider normal perturbations of M compatible with the constraints given by

$$\bar{X}(m, \varepsilon) = X(m) + \varepsilon\varphi(m)N(m),$$

where $m \in M$, $-\varepsilon_0 < \varepsilon < \varepsilon_0$, X and $\bar{X}$ are, respectively, the unperturbed and perturbed position vectors on M, N the Gauss map on M, and φ the perturbation function. The first variation of the energy yields the Euler–Lagrange equation

$$2H - 4C(x^2 + y^2) + \gamma = 0 \quad \text{on } M, \tag{5}$$

where H is the mean curvature function of M, γ is a Lagrange multiplier, and the coordinate functions x and y are given by $X(m) = (x(m), y(m), z(m))$. We also made the assumption at this point that the axis of rotation is the z-axis. The second term in (5) comes from the variation of the moment of inertia; the first one, of course, from the variation of the area.

Let now M be a solution of (5) with centroid at the origin. The second variation of the energy is the following quadratic form:

$$V[\varphi] = \iint_M (|\nabla\varphi|^2 - q\varphi^2) \, dM, \tag{6}$$

where ∇ is the gradient operator on M,

$$q = 4H^2 - 2K - 8C(\xi x + \eta y), \tag{7}$$

and ξ and η are defined by $N(m) = (\xi(m), \eta(m), \zeta(m))$.

An equilibrium configuration, i.e., a solution M of (5), will be called *stable* if $V[\varphi] > 0$, for all nonzero perturbations φ compatible with the constraints (3, 4), i.e., such that

$$\int_M \varphi \, dM = 0, \tag{8}$$

$$\int_M x\varphi \, dM = \int_M y\varphi \, dM = 0. \tag{9}$$

We will also need to consider the linearized operator L associated with the quadratic form V (i.e., the Jacobi operator):

$$L[\varphi] = \Delta\varphi + q\varphi, \tag{10}$$

where Δ is the Laplacian on M.

Everything so far is valid for any open set $\Omega \subset \mathbb{R}^3$ with smooth boundary M, except that the perturbation function φ should in any case have compact support.

3. Stability

We now assume that M is a simply-connected surface of revolution about the z-axis. It is easily seen (e.g., [3]) that M is then compact and symmetric with respect to the xy-plane and that each hemisphere projects simply onto the equatorial plane (see Figure 1).

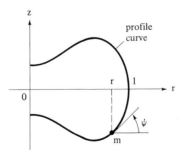

FIGURE 1

Frederic Brulois

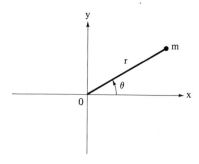

FIGURE 2

Therefore we can use as local coordinates on M the polar coordinates r and θ which are "defined" in Figures 1 and 2. Notice that the equatorial radius has been chosen as scaling length. We shall restrict attention to the lower hemisphere. In those coordinates, the Euler–Lagrange equation has a first integral, namely

$$\sigma(r) = (1 - C)r + Cr^3, \tag{11}$$

where $\sigma = \sin \psi$ and ψ is as indicated in Figure 1. The availability of the explicit first integral (11) is crucial for what follows, since it implies that all coefficients in (6, 7, 8, 9) are explicitly known.

We should be aware, however, that the perturbations φ do not necessarily possess the same equatorial symmetry, but it can be shown that it suffices to consider perturbations with that symmetry. We now sketch a proof of this fact. Let $\varepsilon \mapsto \Omega_\varepsilon$ be a smooth family of admissible perturbations of M with $\partial \Omega_0 = M$. Let $\hat{\Omega}_\varepsilon$ be obtained from Ω_ε by Steiner symmetrization with respect to the xy-plane. Then $\hat{\Omega}_\varepsilon$ and Ω_ε have the same volume, same moment of inertia, and their centroids are at the origin, but

$$\text{area}(\partial \hat{\Omega}_\varepsilon) \leqslant \text{area}(\partial \Omega_\varepsilon);$$

hence

$$\text{energy}(\hat{\Omega}_\varepsilon) \leqslant \text{energy}(\Omega_\varepsilon);$$

finally $\hat{\Omega}_0 = \Omega_0$. Therefore $\varepsilon \mapsto \hat{\Omega}_\varepsilon$ is again a family of admissible perturbations of M but with less energy; and we thus conclude.

We now proceed to do a Fourier decomposition of φ:

$$\varphi(m) = \varphi(r, \theta) = \frac{1}{2}a_0(r) + \sum_{n=1}^{\infty} (a_n(r)\cos n\theta + b_n(r)\sin n\theta).$$

Then the second variation decomposes into a series of quadratic forms V_n:

$$V[\varphi] = \frac{1}{2}V_0[a_0] + \sum_{n=1}^{\infty} (V_n[a_n] + V_n[b_n]),$$

where V_n is given by

$$V_n[f] = \pi \int_0^1 \left[(1 - \sigma^2)f_r^2 + \left(\frac{n^2}{r^2} - q \right)f^2 \right] \frac{r}{\sqrt{1 - \sigma^2}} \, dr, \qquad (12)$$

where σ is given by (11) and q (the function given by (7) above) takes, in the axisymmetric case, the striking form

$$q = \frac{1}{r}(r\sigma \, \sigma_r)_r.$$

The linearized operator L_n associated with V_n is

$$L_n[f] = \frac{\sqrt{1 - \sigma^2}}{r}(r\sqrt{1 - \sigma^2}f_r)_r + \left(q - \frac{n^2}{r^2} \right)f. \qquad (13)$$

Finally, the volume constraint (8) and centroid constraint (9) are, respectively, equivalent to

$$\int_0^1 a_0 \frac{r}{\sqrt{1 - \sigma^2}} \, dr = 0,$$

$$\int_0^1 a_1 \frac{r^2}{\sqrt{1 - \sigma^2}} \, dr = \int_0^1 b_1 \frac{r^2}{\sqrt{1 - \sigma^2}} \, dr = 0.$$

Thus, stability as defined above is equivalent to the conjunction of the following statements: $(0), (1), (2), \ldots, (n), \ldots$.

(0) toroidal stability: $V_0[f] > 0$, for all f such that

$$\int_0^1 f(r) \frac{r}{\sqrt{1 - \sigma^2}} \, dr = 0;$$

(1) shear stability: $V_1[f] > 0$, for all f such that

$$\int_0^1 f(r) \frac{r^2}{\sqrt{1 - \sigma^2}} \, dr = 0;$$

(n) n-lobe stability: $V_n[f] > 0$, for all f.

It is clear from (12) that $V_n[f]$ is an increasing function of n. Hence 2-lobe stability implies n-lobe stability for all $n \geqslant 2$. And stability is thus equivalent to the conjunction of statements (0), (1), and (2) above.

For $C = 0$, the energy functional reduces to the area term and M is a sphere; since, as is well-known, the sphere is stable, we therefore have stability in a neighborhood of $C = 0$ by continuity. Thus the limit of stability C_0 is the smallest value of C for which zero is an eigenvalue of the Jacobi operator L given by (10) and restricted to the linear subspace defined by $(8, 9)$.

An argument, still incomplete at present, will show (it is hoped) that toroidal stability holds for all axisymmetric equilibrium configurations M. The argument is based on the fact that one can guess (!) a solution of $L_0[f] = 0$, namely $f = \sqrt{1 - \sigma^2}$ where σ is given by (11). The proof of shear stability is still incomplete as well. In that regard, the numerical evidence presented by Brown and Scriven [2] indicates that toroidal stability and shear stability do hold for all values of C. We now assume this to be true.

4. n-Lobe Stability

A value of C will be called *characteristic* for n-lobe stability if it is such that zero is an eigenvalue of L_n, more precisely if there exists a non-zero solution f of

$$L_n[f] = 0 \tag{14}$$

which satisfies the boundary conditions

$$f \text{ is bounded at } r = 0, \tag{15}$$

$$f \text{ is smooth at } r = 1. \tag{16}$$

That f must be smooth at $r = 1$ can be seen using arclength instead of r as a parameter on the profile curve (Fig. 1) in a neighborhood of $r = 1$. As seen in the previous section, the limit of stability C_0 is determined by two-lobe stability; in other words, C_0 is the smallest characteristic value of C for two-lobe stability.

Equation (14) is a Fuchsian equation with 8 regular singularities, in particular the coefficients of f, f_r, and f_{rr} are polynomials in r. Because of the symmetry of the equation, the change of variable $x = r^2$ yields again a Fuchsian equation:

$$x^2 P(x) f_{xx} + x Q(x) f_x + R(x) f = 0 \qquad \text{for } 0 < x < 1 \tag{17}$$

and boundary conditions similar to (15, 16):

$$f \text{ is bounded at } x = 0, \tag{18}$$

$$f \text{ is smooth at } x = 1, \tag{19}$$

where

$$P(x) = 4 - 4D^2 x - 8CDx^2 - 4C^2 x^3,$$

$$Q(x) = 4 - 6D^2 x - 16CDx^2 - 10C^2 x^3,$$

$$R(x) = -n^2 + 2D^2 x + 16CDx^2 + 18C^2 x^3, \qquad \text{and } D = 1 - C.$$

The advantage of equation (17) is that it has only 5 singularities. These singularities are regular and their characteristic exponents are obtained as the roots of the indicial equation. We obtain the following Riemann P symbol:

$$f(x) = P \left\{ \begin{array}{ccccc} 0 & 1 & \omega & \bar{\omega} & \infty \\ \frac{n}{2} & 0 & 0 & 0 & 3 \\ -\frac{n}{2} & \frac{1}{2} & \frac{1}{2} & \frac{1}{2} & -\frac{3}{2} \end{array} \quad x \right\}, \tag{20}$$

where

$$\omega = \frac{1}{2} - \frac{1}{C} + i \sqrt{\frac{1}{C} - \frac{1}{4}}.$$

The first row in (20) is a list of the singularities, and the exponents appear below the corresponding singularity. This solution (20) of (17) will satisfy the boundary conditions (18), (19) if and only if f is both a fundamental solution at $x = 0$ belonging to exponent $n/2$ and a fundamental solution at $x = 1$ belonging to exponent 0.

In general, a Fuchsian equation will have a solution which is fundamental at two singularities only for special values of the parameters in the equation. Apart from the case of 3 singularities, namely the hypergeometric equation, no expressions are known for those "characteristic values". There are several methods [9,4], especially for the Heun equation (i.e.: with 4 singularities), which involve some kind of infinite process, like continued fractions, but which did not seem to apply simply in this case. We present instead an ad hoc method.

Let now f represent the fundamental solution of (17) at $x = 0$ belonging to exponent $n/2$. Thus for f to be an eigenfunction belonging to zero it is necessary and sufficient for f to be analytic at $x = 1$. Using the Frobenius method, we get

$$f(x) = x^{n/2} \sum_{m=0}^{\infty} a_m x^m, \tag{21}$$

where the a_m are given by the recurrence relation

$$\begin{cases} a_0 = 1, \quad a_{-1} = a_{-2} = \cdots = 0, \quad \text{and} \\[2mm] a_m = \dfrac{(2m + n)}{(2m)(2m + 2n)}[D^2(2m + n - 3)a_{m-1} \\[2mm] \qquad + 2CD(2m + n - 6)a_{m-2} + C^2(2m + n - 9)a_{m-3}]. \end{cases} \tag{22}$$

The method consists in determining the radius of convergence of (21) using (22). It is based on the fact that the recurrence relation (22) has a strikingly simple structure, which makes it possible to estimate a_m recursively, and the fact that $x = 1$ is the singularity closest to $x = 0$, at least for the values of C that matter, namely $0 < C < 1$. See Figure 3 where the 5 singularities

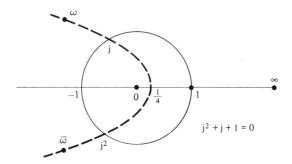

FIGURE 3. x-complex plane.

are represented by heavy dots; as C varies, observe that ω and $\bar{\omega}$ move on the parabola and remain *outside* the unit disc for $0 < C < 1$.

Thus, for most values of C, the radius of convergence of (21) will be 1 and, for the characteristic values of C, it will be $|\omega| > 1$. A proof by induction and a simple estimation of a_m using (22) yields the following.

Theorem 1. *Suppose* $0 < C < 1$, $n \geqslant 0$, *and* a_μ, $a_{\mu+1}$, $a_{\mu+2} > 0$, *for some* $\mu \geqslant \max(0, (3 - n)/2)$. *Then*

(i) $a_m > 0$, *for all* $m \geqslant \mu$,
(ii) $\lim\sup_{m \to \infty} \sqrt{|a_m|} = 1$; *hence in particular,*
(iii) C *is not a characteristic value for n-lobe stability.*

For $n = 3$ and any $C \in (0, 1)$, it turns out that $a_0, a_1, a_2 > 0$, so that Theorem 1 applies and no $C \in (0, 1)$ is a characteristic value for 3-lobe stability. But $C = 1$ is a characteristic value for 3-lobe stability; indeed, for $C = 1$, equation (17) is transformed into a hypergeometric equation by using x^3 as a new independent variable, and one easily obtains the eigenfunction, namely $f = x^3 = r^6$.

Therefore $C = 1$ is the smallest characteristic value for 3-lobe stability. Hence by a continuity argument and the fact that 2-lobe stability implies 3-lobe stability, we obtain $0 < C_0 < 1$.

From here on, set $n = 2$. One sees from (22) that $a_m(C)$ is a polynomial in C. One can show that, for $m \geqslant 2$,

(i) $a_m(C) > 0$, as $C \to 0+$,
(ii) $a_m(C) < 0$, as $C \to 1-$, and
(iii) $a_m(C) = 0$ has a unique solution C_m in the interval $0 < C < 1$.

This, together with Theorem 1, yields the following.

Theorem 2. (i) *If* $\check{C} \in (0, 1)$ *is such that* $a_\mu(\check{C})$, $a_{\mu+1}(\check{C})$, $a_{\mu+2}(\check{C}) > 0$ *for some* $\mu \geqslant 1$, *then* $\check{C} < C_0$.
(ii) *If* $\hat{C} \in (0, 1)$ *is such that* $a_\mu(\hat{C})$, $a_{\mu+1}(\hat{C})$, $a_{\mu+2}(\hat{C}) < 0$ *for some* $\mu \geqslant 1$, *then* $C_0 < \hat{C}$.

The following numerical scheme exploits these facts:

(i) obtain an approximate value of C_m for a large m; e.g. taking $m = 33$, the approximate value $C_{33} \simeq .4488628130$ was obtained in 22 minutes on a Hewlett-Packard 15-C programmable pocket calculator;
(ii) sandwich the "numerical guess" for C_0 obtained in (i) by two numbers $\check{C}$ and $\hat{C}$; e.g. take

$$\check{C} = .448862813 \quad \text{and} \tag{23}$$

$$\hat{C} = .448862814; \tag{24}$$

(iii) check that $a_\mu(\check{C})$, $a_{\mu+1}(\check{C})$, $a_{\mu+2}(\check{C}) > 0$, for μ large enough; e.g. for $\check{C}$ given by (23), one *proves* that these inequalities hold for $\mu = 27$;

(iv) check that $a_\mu(\hat{C})$, $a_{\mu+1}(\hat{C})$, $a_{\mu+2}(\hat{C}) > 0$, for μ large enough; e.g. for $\hat{C}$ given by (24), one proves that these inequalities hold for $\mu = 25$;

(v) conclude from Theorem 2 that $\check{C} < C_0 < \hat{C}$; e.g. we have shown that estimates (2) for C_0 hold.

5. Concluding Remarks

The method used to obtain the characteristic values of the Fuchsian equation (17) appears to be new. It is possible that the method generalizes to the general Fuchsian equation. We hope to return to this question in a later paper.

Brown and Scriven [2] showed numerically that, at $C = 1$, namely the limit of 3-lobe stability, a 3-lobed family of equilibrium configurations bifurcates from the axisymmetric family considered here. Because the eigenfunction is explicitly known ($f = r^6$ as we saw), it appears to be well within reach to prove that fact.

I would like to thank Professor James Ross for getting me interested in this problem [8], Professor Hans Samelson and the American Mathematical Society for publishing a query on Fuchsian equations, and Professor A. Ronveaux of N.D. de la Paix University in Belgium for answering my query and introducing me to the modern literature on Fuchsian equations. Finally, I would like to thank Professors Henry Wente and Josef Bemelmans for helpful conversations.

References

1. P. Appel, *Traité de Mécanique Rationelle*, vol. 4, chap. 9, Paris (1932).

2. R.A. Brown and L.E. Scriven, *The shape and stability of rotating liquid drops*. Proc. R. Soc. London **A371** (1980), 331–357.

3. S. Chandrasekhar, *The stability of rotating liquid drops*. Proc. R. Soc. London **A286** (1965), 1–26.

4. A. Erdelyi, *Fuchsian equation of second order with four singularities* Duke Math. J. **9** (1942), 107–112.

5. J. Plateau, *Recherches expérimentales et théoriques sur les Figures d'équilibre d'une masse liquide sans pesanteur*, Mém. Acad. Roy. Sci. Lett. Belg. **16–37** (1843–69).

6. H. Poincare, *Capillarité*, Paris (1895).

7. Lord Rayleigh, *The equilibrium of revolving liquid under capillary force*, Phil. Mag. **28** (1914), 161–170.

8. J. Ross and F. Brulois, *The stability of axisymmetric rotating drops*, Astérisque. **118** (1984), 219–226.

9. N. Svartholm, *Die lösung der Fuchsschen Differentialgleichung zweiter Ordnung durch hypergeometrische Polynome*, Math. Annalen. **116** (1939), 413–421.

Numerical Methods for Propagating Fronts

James A. Sethian

In many physical problems, a key aspect is the motion of a propagating front separating two components. As fundamental as this may be, the development of a numerical algorithm to track the moving front accurately is difficult. In this report, we describe some previous theoretical and numerical work. We begin with two examples to motivate the problem, followed by some analytical results. These theoretical results are then used as a foundation for two different types of numerical schemes. Finally, we describe the application of one of these schemes to our work in combustion.

I. Motivation

A. Crystal Growth

A relatively straightforward example is the growth of a solid immersed in a supercooled liquid, discussed extensively in Langer [9]. To illustrate, we imagine a solid ice crystal placed in a bath of water that has been supercooled below its freezing point. We limit the discussion to two dimensions and neglect a variety of effects such as convective heat transport, solid impurities, and crystalline anisotropy. Let $\gamma(t)$ be the closed curve representing the boundary between ice and water at time t, with ice inside the curve.

The diffusion equation for the temperature T holds both inside and outside $\gamma(t)$, namely, $\partial T_{s(l)}/\partial t = c_{s(l)} \nabla^2 T_{s(l)}$, where ∇^2 is the two-dimensional Laplacian, $c_{s(l)}$ is the thermal diffusion coefficient, and $T_{s(l)}$ is the temperature in the solid(s)/liquid(l) region. Conservation of heat flux across the boundary interface must include the heat required to go from solid to liquid, thus $c_s \partial T_s/\partial n - c_l \partial T_l/\partial n = L V_\gamma$, where $\partial/\partial n$ is differentiation in the normal direction, L is the latent heat of formation, and V_γ is the velocity of the boundary

$\gamma(t)$ along its normal vector field. Finally, the thermodynamic boundary condition, which includes the effects of surface tension, is given by the Gibbs–Thomson relation (see Turnbull [23]). At each point $\bar{x}$ of the boundary $\gamma(t)$, we require that

$$T(\bar{x}) = T_M(1 - \varepsilon K(\bar{x})),$$

where T_M is the melting temperature, ε is a constant, and $K(\bar{x})$ is the curvature of the boundary $\gamma(t)$ at $\bar{x}$.

Thus, if points of negative curvature are concave towards the solid region, they yield a higher temperature than those with positive curvature, and this variation in the solidification rate along the boundary as a function of curvature provides intricate growth patterns. For further information, see [9], [10], [13], [14], [16], [20].

B. Flame Propagation

Much work surrounding the stability/instability of a flame is aimed at under standing the "turbulization" or wrinkling of a flame front and its interaction with the hydrodynamic flow field. The pioneering work in this field is the analysis of a plane flame front by Landau [8]. The flame front is idealized as a surface of discontinuity, i.e., a closed curve $\gamma(t)$, separating regions of constant steady-state velocity, density, and temperature. In Landau's model, the flame speed V_γ of the curve along its normal vector field is constant. By ignoring all but hydrodynamic effects, flames are shown to be unstable to perturbations in velocity and pressure around a mean state. Since this conclusion is physically unreasonable, Markstein [11] postulated that the flame speed depended on the curvature so that

$$V_\gamma = V^0(1 - \varepsilon K(\gamma)),$$

where V^0 is the constant speed of a flat flame, ε is a constant, and $K(\gamma)$ is the curvature. The motivation behind such an assumption, loosely speaking, is that parts of the flame front which bend in towards the hot burnt region are subject to greater heat and hence burn faster; slower flame speeds are thus implied for fingers reaching out into cool gas.

Using linear stability analysis of this model, Markstein demonstrated the stabilizing effect of curvature. Since then, there have been numerous investigations of flame stability for a variety of combustion models. A comprehensive though now outdated account may be found in Markstein [12], here, we also mention the work of Sivashinsky [22], Frankel and Sivashinsky [4], and Zeldovich [24].

II. Theoretical Results

Starting with a simple, smooth, closed initial curve $\gamma(0)$ in R^2, let $\gamma(t)$ be the one parameter family of curves, where $t \in [0, \infty)$ is time, generated by moving the initial curve along its normal vector field with speed F a given function of the curvature. Let $\vec{x}(s, t)$ be the position vector which parameterizes $\gamma(t)$ by s, $0 \leqslant s \leqslant S$, $\vec{x}(0, t) = \vec{x}(S, t)$. The curve is parameterized so that the interior is on the left in the direction of increasing s. With $K(s, t)$ as the curvature at $\vec{x}(s, t)$, the equations of motion are

$$\vec{n}(s, t) \cdot \frac{\partial \vec{x}(s, t)}{\partial t} = F(K(s, t)) \tag{1}$$

$$\vec{x}(s, 0) = \gamma(0) \quad \text{prescribed;} \quad s \in [0, S] \quad t \in [0, \infty),$$

where $\vec{n}(s, t)$ is the unit normal vector at $\vec{x}(s, t)$. If $\alpha(s)$ corresponds to arclength, then $d\alpha = g(s, t)\, ds$ where $g(s, t) = (x_s^2 + y_s^2)^{1/2}$. From the above, one can produce an evolution equation for the metric g, namely

$$g_t(s, t) = g(s, t)K(s, t)F(K(s, t)), \tag{2}$$

and an evolution equation for the curvature K, namely

$$K_t(s, t) = -[F_s(K(s, t))g^{-1}(s, t)]_s g^{-1}(s, t) - K^2(s, t)F(K(s, t)). \tag{3}$$

We point out two facts: (1) The case $F(K) = K$ occurs in the modeling of grain boundaries in metals, and has been studied extensively in Brakke [1], and (2) for $F(K) = K$, a recent result by Huisken shows that convex surfaces remain smooth as they collapse; the smoothness of non-convex surfaces remains an open question.

Let $\text{Var}(t)$ be the total variation of the front at time t, defined as

$$\text{Var}(t) = \int_0^S |K(s, t)| g(s, t)\, ds. \tag{4}$$

Then we have the following (see [19]).

Proposition. *Consider a front moving with speed $F(K)$, as in equation (1). Assume that $\gamma(0)$ is non-convex, so that $K(s, 0)$ changes sign, and assume K is zero at a finite number of points. Assume that F is twice differentiable, and that $K(s, t)$ is twice differentiable for $0 \leqslant s \leqslant S$ and $0 \leqslant t \leqslant T$. Then:*

(1) if $F_K(0) \leqslant 0 \ (F_K(0) \geqslant 0)$, then

$$\frac{d\,\text{Var}(t)}{dt} \leqslant 0 \quad \left(\frac{d\,\text{Var}(t)}{dt} \geqslant 0 \right),$$

(2) if $F_K(0) < 0 \ (F_K(0) > 0)$ and $K_s(0) \neq 0$, then

$$\frac{d\,\text{Var}(t)}{dt} < 0 \quad \left(\frac{d\,\text{Var}(t)}{dt} > 0 \right),$$

for $0 \leqslant t \leqslant T$.

Remarks. This proposition states that if $F_K < 0$ wherever $K = 0$, then the front "flattens out". If $\gamma(t)$ is convex, the proposition is trivial, since $\text{Var}(t) = \int_0^S Kg\, ds = 2\pi$. If $F_K < 0$, the "energy" of the front dissipates. We also have the following, which applies to a front moving at constant speed.

Corollary. *If $F_K = 0$, then the total variation is constant.*

We focus on the case $F(K) = 1 - \varepsilon K$, where ε is a constant. For $\varepsilon > 0$, $d\operatorname{Var}(t)/dt < 0$ and the energy dissipates; for $\varepsilon = 0$, $d\operatorname{Var}(t)/dt = 0$ and the energy is constant. We may rewrite the curvature evolution equation as

$$K_t = \varepsilon K_{\alpha\alpha} + \varepsilon K^3 - K^2, \tag{5}$$

where here we have changed variables and taken the derivative of curvature with respect to arclength to eliminate the metric g. This is a reaction-diffusion equation, in which the reaction term $(\varepsilon K^3 - K^2)$ is mitigated by the diffusion term $(\varepsilon K_{\alpha\alpha})$. With $\varepsilon > 0$, it can be shown that the solution stays smooth for all time. With $\varepsilon = 0$, we have $K_t = -K^2$ which is singular at finite t if the initial curvature is anywhere negative.

This situation is reminiscent of the development of shocks in hyperbolic conservation laws. Consider Burgers equation with viscosity, namely

$$u_t + uu_x = \varepsilon u_{xx}. \tag{6}$$

It is well-known (see [5, 6]), that for $\varepsilon > 0$, the "viscosity" term produces smooth solutions and the energy dissipates smoothly. For $\varepsilon = 0$, shock discontinuities can develop in the solution, even for smooth initial data, and an "entropy" condition is used to provide a weak solution beyond breakdown. Loosely put, this entropy condition requires that characteristics "enter" shock discontinuities, and the weak solution obtained is the limiting solution of Equation (6) as $\varepsilon \to 0$.

In fact, it is easy to see that our equations for a propagating curve can be recast as a single conservation law with viscosity. Suppose we write the propagating curve as a function, that is, let $y = \phi(x, t)$ be the height y of the moving curve at a point x at time t, drawn in the $x - y$ plane. Then the equation of motion for a front moving with speed $F(K) = 1 - \varepsilon K$ can be written as

$$\phi_t = \left(1 + \varepsilon \frac{\phi_{xx}}{(1 + \phi_x^2)^{3/2}}\right)(1 + \phi_x^2)^{1/2}. \tag{7}$$

Let $u = \phi_x$. Then

$$u_t - [G(u)]_x = \varepsilon[u_x/((G[u])^2)]_x, \tag{8}$$

where $G[u] = (1 + u^2)^{1/2}$. Thus, the equation for the propagating tangent ϕ_x satisfies a Burgers-type equation with viscosity.

What role does the entropy condition from the shock case play in the propagating curve? Imagine the front as a flame separating a burnt interior from an unburnt exterior; each point is transformed from unburnt to burnt when touched by the propagating front. The entropy condition, loosely speaking, may be stated as follows: Once a particle is burned, it stays burnt. The position of the front at a time t may be obtained by using Huygens principle: the front is formed by the envelope of all disks of radius t centered along the initial curve. This solution will be called the "Huygens principle construction"

A different way to look at what is happening is to construct the normals to initial curve, which we shall call ignition curves. Consider the following construction: Starting with the initial curve, allow the front to burn along the ignition curves until there is a collision. Eliminate the curves (characteristics) that collide, and continue moving the front along the remaining curves. As time progresses, the front will consist of a shrinking subset of the original curve, and any point on the solution can be traced back to the initial curve. Conversely, there will be points along the initial curve that do not affect the solution beyond some time, and thus the solution becomes irreversible. Thus, characteristics always trace back to the initial line. Furthermore, $d\operatorname{Var}/dt \leqslant 0$, since $\operatorname{Var}(t)$ is the integral of a positive quantity over a set whose length is a non-increasing function of time (here, one evaluates $\operatorname{Var}(t)$ over open intervals where the curvature is defined together with jumps in the tangent at the corners).

These two constructions yield the same solution. The removal of ignition curves in the propagating flame case corresponds to the elimination of colliding characteristics in the shock case. Furthermore, this weak solution can be shown to be the limiting case of the smooth solution with viscosity as $\varepsilon \rightarrow 0$.

The above suggests at least two numerical schemes for following fronts. Most algorithms place marker particles along the front and advance the position of the particles in accordance with a set of finite difference approximations to the equations of motion. Such schemes usually go unstable and blow up as the curvature builds around a cusp, since small errors in the position produce large errors in the determination of the curvature. One alternative is to consider the reformulation equations of motion as a conservation law with viscosity and solve these equations with the techniques developed for gas dynamics. These techniques, based on high-order upwind formulations, are particularly attractive, since they are highly stable, accurate, and preserve monotonicity. We have made some preliminary tests of such schemes applied to our problem of propagating fronts in crystals and flames, with extremely encouraging results, and will report on this work elsewhere [17].

Another alternative is to rely on the Huygens principle idea and follow the motion of the *interior* of the "burnt" region, rather than the boundary. These techniques are of the "volume of fluid" type, and it is those we shall discuss and demonstrate below.

III. Numerical Methods Based on Volume of Fluid Techniques

Here, we briefly describe an application of a volume of fluid method (SLIC) for the case $\varepsilon = 0$. For details, see [3], [15], [17].

We lay down a square grid i, j of uniform mesh size on the computational domain, and assign a number f_{ij}, $0 \leqslant f_{ij} \leqslant 1$ corresponding to the fraction of burnt fluid within cell i, j. Each cell's material is moved in the direction $\vec{u} = (u, v)$ by executing a fractional step in each direction. An interface is drawn in the cell representing the boundary of the material, where the orientation of the interface depends on the value of f_{ij} and the f_{ij}'s in the cell's neighbors. This interface is then transported in the x direction a distance $u\Delta t$, and the process is then repeated for the sweep in the y direction, providing new f_{ij}.

This algorithm can be used to advance a front along its normal vector field using Huygens principle. For the moment, assume that $F(K) = 1$ and consider L angles, $\Theta_l = 2\pi(l - 1)/L$, $l = 1, \ldots, L$. Given any cell with volume fraction f_{ij}^n, the material in that cell is moved a distance Δt in each of the l directions $(\cos \Theta_l, \sin \Theta_l)$. As $L \to \infty$ and the mesh size goes to zero; this corresponds to drawing a disk of unit radius around the center of the cell. The Huygens principle construction says that the envelope formed by all such disks (that is, for all i, j) gives the front advanced a unit distance along its normal field. Thus, let $f_{ij}^n \Theta_l$ be the array of volume fractions obtained by moving the fractions in the direction Θ_l and let $f_{ij}^n \Theta_0 = f_{ij}^n$. The new volume fractions approximating the front advanced one time step will be given by

$$f_{ij}^{n+1} = \max_{0 \leqslant l \leqslant L} f_{ij}^n \Theta_l.$$

Since f_{ij} is often either zero or one, careful programming will limit the computing effort to the boundary. The generalization to three dimensions is straightforward.

This numerical method filters out high wavelengths and smooths the solution by limiting the oscillations of the front to the order of one cell width. The method can be extended to a front moving with speed $F(K)$ by determining the curvature from the volume fractions. A method based on osculating circles has been developed [2]; there, the cell size determines the smallest possible osculating circle and hence bounds the curvature and smooths the solution. In spite of this smoothing, such a curvature algorithm can provide a valuable tool for analyzing a moving front, since the entropy condition naturally generates weak solutions and the mesh size can be systematically refined to allow larger curvatures in a controlled way. Thus, if a mesh size h is used with maximum allowable curvature K_h, the algorithm produces a weak solution with this bound; by refining the mesh size, one can investigate both possible blow up in the curvature and the nature of the solution beyond the singularity.

IV. Examples

To begin, we use the above algorithm to follow a kidney bean shape expanding with constant velocity $F(K) = .5$. In Figure 1A (see [18]), we show the exact solution, found by solving the equations of motion, for a burnt region initially bounded by four joined semi-circles. The position of the front is shown for various values of t. The concave region sharpens into a corner at $t = 1$, which then opens up and smooths out. In Figure 1B, we show the results of the

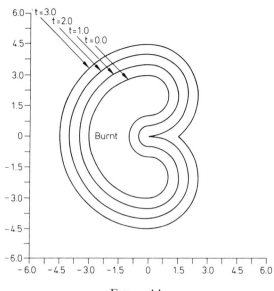

FIGURE 1A

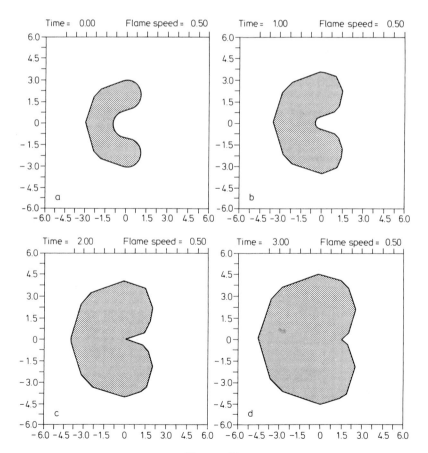

FIGURE 1B

application of the Huygens principle algorithm to this problem, with 8 angles and a 60 × 60 mesh of cells. The numerical method follows the development of the corner, and the ensuing weak solution.

Next, we use this algorithm to analyze the competing effects of viscosity and exothermicity on flame propagation (see [20, 21]). The setup of the problem is as follows. A fluid swirling in a closed box is ignited halfway up the left wall. The motion of the flame consists of two effects: it burns in a direction normal to itself at a constant speed, and it is advected by the underlying velocity field. Four different experiments were performed. First (A), we study inviscid, constant density flow. In this experiment, the flow is

FIGURE 2. Swirling fluid.

potential (no-slip condition is ignored), and the flame is cold (no expansion upon combustion), thus the flame simply burns and is passively advected by the swirling flow. In the second experiment (B), inviscid flow with exothermicity, the flow is potential but the fluid expands as it burns, thus the pressure in the vessel is increased, changing the hydrodynamic field. In the third experiment (C), we have viscous flow with a cold flame. Here, the no-slip condition on the walls is satisfied, creating counterrotating corner eddies which contort and stretch the flame. Finally, in the last experiment, we have viscous flow with exothermicity.

In the two viscous runs, the flow was started two seconds before ignition so that recirculation zones would have time to develop. The results are shown in Figure 2A–2D. In the inviscid, constant density case, the flame is smoothly advected by the large vortex in the center. In the inviscid, exothermic case, the velocity field produced by volume expansion and the rise in pressure and flame speed cause the flame to spiral in towards the center at a faster rate. In the viscous, constant density case, the flame front is twisted by the eddies that develop in the corners; the flame is carried over the eddies and dragged backwards into the corners. The effect of these eddies is to extend the length of the flame front, bringing it into contact with unburnt fuel and increasing the rate at which the vessel becomes fully burnt. Finally, when both viscous and exothermic effects are combined, the flame is both wrinkled due to the turbulence of the flow (hence increasing the surface area of the flame) and carried by the exothermic field; in addition both the flame speed and pressure increase. The effect of these factors is to greatly decrease the amount of time required for complete conversion of reactants to products.

Acknowledgments. This work was supported in part by the Applied Mathematical Sciences subprogram of the Office of Energy Research, U.S. Department of Energy under contract DE-AC03-76SF00098.

References

1. K.A. Brakke, *The motion of a surface by its mean curvature*, Princeton University Press, Princeton, New Jersey (1978).
2. A.J. Chorin, *Curvature and solidification*, J. Comp. Phys. (to appear).
3. A.J. Chorin, *Flame advection and propagation algorithms*, J. Comp. Phys. **35** (1980), 1–11.
4. M.L. Frankel and G.I. Sivashinsky, *The effect of viscosity on hydrodynamic stability of a plane flame front*, Comb. Sci. Tech. **29** (1982), 207–224.
5. P.R. Garabedian, *Partial differential equations.* John Wiley and Sons, New York (1961).
6. E. Hopf, *The partial differential equation $u_t + uu_x = \mu u_{xx}$*, Comm. Pure Appl. Math. **3** (1950), 201.
7. G. Huisken, *Flow by mean curvature of convex surfaces into spheres*, Preprint (1984).
8. L. Landau, *On the theory of slow combustion*, ACTA Physiocochimica, URSS **19** (1944), 77–85.

9. J.S. Langer, *Instabilities and pattern formation in crystal growth*, Rev. Mod. Phys. **52** (1980), 1–28.

10. J.S. Langer and H. Muller–Krumhaar, *Mode selection in a dendritelike nonlinear system*, Phys. Rev. A **27** (1983), 499–514.

11. G.H. Markstein, *Experimental and theoretical studies of flame front stability*, J. Aero. Sci. **18** (1951), 199–209.

12. G.H. Markstein, *Non-Steady flame propagation*, Pergamon Press, MacMillan Company, New York (1964).

13. W.W. Mullins and R.F. Sekerka, J. Appl. Phys. **34** (1963), 2885.

14. F.A. Nichols and W.W. Mullins, Trans. Met. Soc., AIME **223** (1965), 1840.

15. W. Noh and P. Woodward, *A simple line interface calculation*, Proceedings, Fifth International Conference on Fluid Dynamics, A.I. van de Vooran and P.J. Zandberger, Eds. Springer-Verlag, New York (1976).

16. B.R. Pamplin, *Crystal Growth*. Pergamon Press, New York (1975).

17. J.A. Sethian, *An algorithm for propagating fronts based on high-order upwind schemes* (in progess).

18. J.A. Sethian, *An analysis of flame propagation*, PhD. Dissertation, University of California, Berkeley, California, June 1982; CPAM Rep. 79.

19. J.A. Sethian, *Curvature and the evolution of fronts*, Communications of Mathematical Physics **101** (1985), 487–499.

20. J.A. Sethian, *Numerical Simulation of Flame Propagation in a Closed Vessel*, Notes on Numerical Fluid Mechanics, Editors: Pandolfi, M. and Piva, R., Proceedings of the Fifth GAMM-Conference on Numerical Methods in Fluid Mechanics, Rome, October 5–7, 1983, Friedr, Vieweg & Sohn, Braunschweig/Wiesbaden, 1984, pp. 324–331.

21. J.A. Sethian, *Turbulent combustion in open and closed vessels*. J. Comp. Phys. **54** (1984), 425–456.

22. G.I. Sivashinsky, *Nonlinear analysis of hydrodynamic instability in laminar flames, I*. Acta Astronautica **4** (1977), 1177–1206.

23. D. Turnbull, *"Phase changes" in solid state physics*, **3**, ed. Seitz, F. and Turnbull, D., Academic Press, New York (1956).

24. Y.B. Zeldovich, *Structure and stability of steady laminar flame at moderately large Reynolds numbers*, Comb. Flame **40** (1981), 225–234.

A Dynamic Free Surface Deformation Driven by Anisotropic Interfacial Forces

Daniel Zinemanas and Avinoam Nir

The dynamic deformation of a viscous droplet under the influence of anisotropic surface forces is examined. Asymptotic calculations are compared with the case of isotropic surface tension with possible relevance to the cleavage stage in the division of biological cells.

Introduction

Consider a viscous fluid droplet suspended in an immiscible infinite viscous fluid domain. The interface between the two regions is a thin layer containing a dispersion of a surface active substance (SAS). The form of the individual particle of the SAS is a slender rod-like filament which can exercise contractile forces along its axis. The filaments are confined to the surface layer with their axis always tangent to the interface. Their motion and orientation are determined by the kinematics of the interface and the velocity gradients there. The combination of the individual contractile forces of the individual SAS particles comprises the major component of the tension in the interface.

A biological example of a system to which the above description was used as a mechanical simulation is cytokinesis, i.e., the division of the cytoplasm and plasma membrane, which is the ultimate step in the process of cell cleavage. Even though much work has been done on this subject, the mechanisms controlling this process are still poorly understood (see reviews by Rappaport [1], Schroeder [2]). However, it is agreed that the driving force is provided by a contractile ring (CR) composed of muscle-like filaments aggregated in the equatorial region of the cell cortex which contracts until division is attained. At the onset of cytokinesis the filaments are uniformly distributed on the cortex and are randomly oriented [10]. The formation of this CR is not well understood; however, Greenspan [5] has pointed out that hydro-dynamic effects could play an important role during this process.

Based on early experiments by Spek [3], Greenspan [4], [5] showed that a drop exposed to a lowering of the surface tension in the polar regions can undergo a similar deformation to that shown by the living cell. Surface movements then, will tend to concentrate the filaments in the equatorial region, forming thus the contractile ring and increasing the tension in that region. This process can become unstable and continue indefinitely once it is triggered.

Calculation based on isotropic surface tension failed, however, to confirm the above hypothesis (Greenspan [4], Sapir and Nir [6]) since the negative curvature developed in the equator was always dominant in stopping the deformation dynamics far from cleavage.

The purpose of this communication is to examine quantitatively the idea that the reorientation of filaments due to surface movement leads to an arrangement similar to the form of the contractile ring, thereby rendering the surface tension anisotropic. The dynamically developed anisotropy can over-come the cleavage inhibition observed in the case of isotropic tension due to the negative curvature at the equator. The idea was used by White and Borisy [7] without a rigorous formulation to show interesting possible surface de-formations.

We shall formulate the general problem for a viscous droplet deformation driven by an anisotropic surface tension, which in turn depends on the local concentration and the orientation distribution of tensile filaments. A simple example related to the biological cell division process will be examined asymptotically.

Formulation

A droplet B having a viscosity μ with surface ∂B is surrounded by an infinite region B^* filled with fluid having a viscosity μ^*. We shall assume that under isothermal conditions the physical properties of the fluids in the domains are constant. Neglecting inertia and gravity effects, the equations of motion in B and B^* are of the form

$$\nabla \cdot \underline{\underline{\sigma}} = 0, \qquad \nabla \cdot \underline{v} = 0 \tag{1}$$

with

$$\underline{\underline{\sigma}} = -P\underline{\underline{I}} + \mu(\nabla \underline{v} + \nabla^\dagger \underline{v}), \tag{2}$$

where μ^* replaces μ in B^*. In equations (1) and (2), $\underline{v}$, $\underline{\underline{\sigma}}$, and P denote the velocity, stress, and pressure fields, respectively.

The interfacial tension is generally anisotropic and therefore is expressed as a tensor $\underline{\underline{\gamma}}$. The extension of such surface properties into B and B^* is not considered here. We shall merely assume that the gradients of $\underline{\underline{\gamma}}$ normal to ∂B

vanish and that $\underline{\gamma}$ has non-trivial components only tangent to ∂B. Hence, the interfacial conditions are continuity of velocities

$$\Delta\underline{v} = 0 \qquad \text{on } \partial B \tag{3}$$

and continuity of surface tractions

$$\Delta\underline{\underline{\sigma}}\cdot\underline{n} = -\nabla\cdot\underline{\underline{\gamma}} + (\underline{\underline{\gamma}}:\nabla\underline{n})\underline{n} \qquad \text{on } \partial B, \tag{4}$$

where Δ denotes a difference across ∂B (outer minus inner) and $\underline{n}$ is a unit vector on ∂B pointing into B^*. The surface shape is given by the function $\phi(x, t) = 0$, implying that n is defined by $\underline{n} = \nabla\phi/|\nabla\phi|$ and the deformation is described by the kinematic condition

$$-\frac{1}{|\nabla\phi|}\frac{\partial\phi}{\partial t} = \underline{v}\cdot\underline{n}. \tag{5}$$

The anisotropic surface tension γ is a function of the concentration of the SAS elements, c, and of the local orientation distribution function of these filaments, $N(\underline{p})$, where $\underline{p}$ is a director along the axis of a SAS particle which is assumed to be a rigid slender body. γ also involves a component γ_m, representing the passive reaction of the cortex region to the filaments' action

$$\underline{\underline{\gamma}} = \underline{\underline{\gamma}}_p(c, N(\underline{p})) + \underline{\underline{\gamma}}_m. \tag{6}$$

Since the filaments are confined to ∂B, a mass concentration equation is of the form

$$\frac{\partial c}{\partial t} + \nabla_s\cdot(\underline{v}_s c) = D_T\nabla_s^2 c + R \qquad \text{on } \partial B, \tag{7}$$

where c is the surface concentration of SAS filaments, R is their rate of production, D_T is a surface translational diffusion coefficient, $\underline{v}_s$ is the velocity field on ∂B, and ∇_s are taken tangent to the interface. Equation (7) is subject to an initial concentration distribution and to the requirement that the filaments are confined to ∂B.

The kinematics of an individual SAS particle is assumed to follow the local translation and rotation of fluid particles in the interface except for possible deviations due to translational and rotational diffusion. Thus, adding a reaction term to the Burgers equation [8], the balance of the orientation distribution function, $N(\underline{p})$, becomes

$$\frac{\partial N}{\partial t} + \underline{v}_s\cdot\nabla_s N = \nabla_p(\underline{\dot{p}}N) + D_R\nabla_p^2 N + \frac{R}{c}(N^* - N) + \frac{D_T}{c}(\nabla_s N\cdot\nabla_s c) \qquad \text{on } \partial B \tag{8}$$

subject to an initial distribution.

Here, D_R is the rotational diffusion coefficient, ∇_p is an operator with respect to $\underline{p}$, and N^* is the orientation distribution function of SAS particles produced by reaction at the rate R. Since the SAS particles are small enough, the local velocity field on ∂B, $\underline{v}_s$, can be considered linear, i.e., $\underline{v}_s = \underline{v}_0 + \nabla_s\underline{v}\cdot\underline{x}$, so that rotational velocity of a particle is

$$\dot{\underline{p}} = \underline{p} \times \nabla_s \underline{v}_s \cdot \underline{p} \times \underline{p}. \tag{9}$$

A general solution of the problem described by equations (1)–(9) requires a cumbersome numerical approach. Some interesting features and behavior can be obtained employing a simplified asymptotic solution. These are demonstrated in the following section for the case of the dividing cell.

Perturbation Solution

Assume a small initial surface tension gradient formed as a result of an initial distribution of active SAS filaments on ∂B. All dependent variables can be expanded in the form

$$g = g^{(0)} + \varepsilon g^{(1)} + 0(\varepsilon^2), \tag{10}$$

with $\varepsilon = a|\nabla \cdot \underline{\underline{\gamma}}|_0/\gamma_0$, where γ_0 is a typical interfacial tension, a is the drop radius, and $|\nabla \cdot \underline{\underline{\gamma}}|_0$ is a typical small initial gradient. Without loss of generality, variables are rendered dimensionless using the radius a, the concentration c_0, the viscosities μ, μ^*, and the mean initial tension, i.e., $\gamma_0 = \frac{1}{2} \text{tr} \, \gamma^{(0)}$.

The shape of the deforming droplet is given by

$$\phi = r - 1 - \varepsilon f^{(1)}(\theta, t) + 0(\varepsilon^2) = 0 \tag{11}$$

while the concentration distribution has the form

$$c(\theta, t) = 1 + \varepsilon c^{(1)}(\theta, t) + 0(\varepsilon^2). \tag{12}$$

Here and in (11), r and θ are radial and azimuthal spherical coordinates originated in the drop center. We shall assume axial symmetry in the boundary and initial conditions and therefore in the entire emerging solution.

Consider an initial concentration profile

$$c^{(1)}(\theta, 0) = \sum_{n=0}^{\infty} C_n P_n(\cos \theta) \tag{13}$$

with a uniform initial distribution function $N^{(0)} = 2/\pi$ everywhere on ∂B. P_n are Legendre polynomials. As time progresses, N becomes also a function of position, θ, and an orientation angle, α, measured with respect to the direction of the surface velocity.

If each surfactant particle exercises a force, k, in the direction of its long axis, the explicit form of equation (10) is then given by

$$\underline{\underline{\gamma}} = \begin{bmatrix} \gamma_{\theta\theta} & 0 \\ 0 & \gamma_{\phi\phi} \end{bmatrix} = c \begin{bmatrix} \int_0^{\pi/2} N \cos \alpha \, d\alpha & 0 \\ 0 & \int_0^{\pi/2} N \sin \alpha \, d\alpha \end{bmatrix} + \underline{\underline{\gamma}}_m \tag{14}$$

with the dimensions of the tension given by $\gamma_0 = kC_0$.

The passive tension, γ_m, which arises in the cortical interfacial layer due to surface viscoelastic deformations is considered a secondary effect and will be neglected here when compared with the components of tension exerted by the active filaments.

The zero order approximation in this case gives the trivial solution of an undeformed static drop with constant concentration, and uniform orientation distribution function of the surfactant on the surface. The normal force balance condition on the surface becomes

$$-\Delta P^{(0)} = \operatorname{tr} \gamma^{(0)} = 2 \tag{15}$$

since $\gamma_{\theta\theta}^{(0)} = \gamma_{\phi\phi}^{(0)} = 1$ when N is uniform. It is also interesting to note here that the solution of the shape of a static drop under constant but anisotropic surface tension such that $\gamma_{\theta\theta} \neq \gamma_{\phi\phi}$, given by Young Laplace's equation, is still a sphere with internal pressure corresponding to an equivalent isotropic surface tension of $\frac{1}{2}\operatorname{tr}\gamma$.

The first order approximation leads to the following set of equations:

$$\nabla \cdot \underline{v}^{(1)} = 0 \qquad \text{in } B \text{ and } B^* $$
$$\nabla P^{(1)} = \nabla^2 \underline{v}^{(1)} \qquad \text{in } B \text{ and } B^* \tag{16}$$

subject to the boundary conditions

$$\underline{v}_B^{(1)} - \lambda \underline{v}_{B^*}^{(1)} = 0, \qquad \lambda = \mu/\mu^* \tag{17}$$

$$\Delta\sigma_{rr}^{(1)} = \operatorname{tr}\gamma^{(1)} - f^{(1)}\operatorname{tr}\gamma^{(0)} - \frac{\partial^2 f^{(1)}}{\partial\theta^2}\gamma_{\theta\theta}^{(0)} - \frac{\partial f^{(1)}}{\partial\theta}\gamma_{\phi\phi}^{(0)}\cot\theta \tag{18}$$

and

$$\Delta\sigma_{r\theta}^{(1)} = -\frac{\partial\gamma_{\theta\theta}^{(1)}}{\partial\theta}. \tag{19}$$

To this order of approximation, neglecting any diffusion of SAS particles and production due to chemical reaction, the concentration and distribution function balances become

$$\frac{\partial c^{(1)}}{\partial t} = -\frac{1}{\sin\theta}\frac{\partial}{\partial\theta}(v_\theta^{(1)}\sin\theta) - 2v_r^{(1)} \tag{20}$$

and

$$\frac{\partial N^{(1)}}{\partial t} = N^{(0)}\cos 2\alpha \frac{\partial v_\theta^{(1)}}{\partial\theta}, \tag{21}$$

while the kinematic condition (5) is

$$\frac{\partial f^{(1)}}{\partial t} = v_r^{(1)} \tag{22}$$

with $\underline{\gamma}^{(1)}$ given by

$$\underline{\gamma}^{(1)} = N^{(0)}\left\{ c^{(1)}\underline{I} + \begin{bmatrix} 1/3 & 0 \\ 0 & -1/3 \end{bmatrix} \int_0^t \frac{\partial v_\theta^{(1)}}{\partial\theta}\,dt \right\}. \tag{23}$$

Equations (17)–(23) are evaluated at $r = 1$. Following Cox [9], the solution to the entire set of equations is expressed as an expansion of spherical harmonics. We have chosen the most simple form of concentration perturbation in equation (13):

$$c^{(1)}(\theta, 0) = C_2 P_2(\cos \theta), \tag{24}$$

which reflects an initial reduction of the SAS concentration at the droplet poles. This choice also yields a solution with fore-and-aft symmetry. The results are presented and discussed below.

Results and Discussion

The orientation distribution function, the tensions on the surface, and the drop shape change in time were calculated and plotted in Figures 1, 2, and 3. Calculations were made for $\varepsilon = 0.05$. The orientation distribution function is

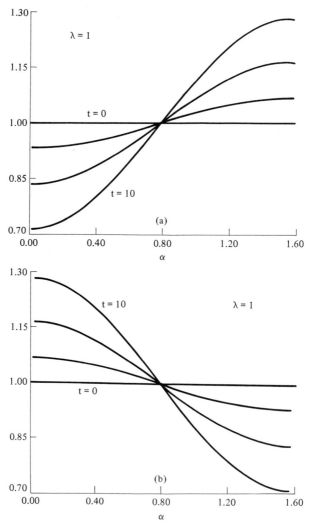

FIGURE 1. The evolution of the orientation distribution function in time. (a) At the equator; (b) at the poles.

directly determined by local surface velocity gradients, as is clear from equation (21). There are two stagnation regions on the interface, at the poles and at the equator. Evidently, the velocity must attain a maximum at some intermediate location, thus the velocity gradient is positive approaching this maximum and negative beyond it. In the region where filaments experience a positive gradient they tend to arrange themselves in the direction of flow, while negative gradients will direct their axis to an orientation perpendicular to the flow. These re-orientations are depicted in Figure 1, where near the equator (Fig. 1a) a tendency to arrange the SAS particles parallel to the equatorial

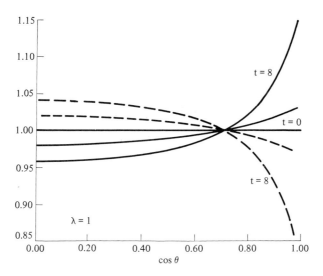

FIGURE 2. The relative change in the components of the surface tension. —— $\gamma_{\theta\theta}$; – – – $\gamma_{\phi\phi}$.

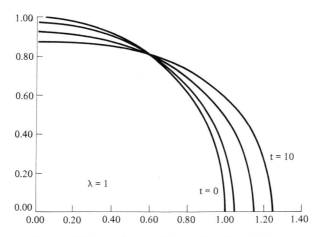

FIGURE 3. The evolution of the surface shape in time.

plane is evident, while near the poles (Fig. 1b) they become parallel to meridional lines. This is exactly the manner in which muscle-like filaments rearrange in the cell cortex to form the equatorial contractile ring which is believed to cause the cell division [10].

The anisotropy due to change in the orientation distribution function which develops during the surface motion is manifested in the form of an anisotropic surface tension. The dynamic development of the tension components is shown in Figure 2. Initially, these components develop differently near the equator and near the poles. Gradients of stabilizing longitudinal tensions appear, which counterbalance the destabilizing gradients in tension due to the concentration profile. This, however, is not sufficient to halt the dynamic deformation of the droplet shape at this order of approximation.

It is anticipated that as time develops, the reduction of longitudinal components and the increase of transverse components of tension near the equator will carry the deformation beyond the point in which it ceased, due to negative curvatures, in the case of isotropic tension [6].

The shapes of the deformed drop are given in Figure 3. It is expected that with higher approximations in the asymptotic solution or using a full numerical scheme, larger deformations of the surface would be obtained, conceivably to a complete cleavage.

Acknowledgments. This work was supported in part by the Fund for Research Advancement at the Technion. D.Z. would also like to acknowledge a fellowship from the Wolf Foundation.

References

[1] G.W. Conrad and R. Rappaport, in: *Mitosis/Cytokinesis*, A.M. Zimmerman and A. Forer, Eds., Academic Press (1981), p. 365.

[2] T.E. Schroeder, in: *Cytoskeletal Elements and Plasma Membrane Organization*, G. Poste and G.K. Nicolson, Eds., Elsevier (1981), p. 169.

[3] J. Spek, Arch Entwicklungs Mech. **44** (1918), 5–113.

[4] H.P. Greenspan, Stu. Appl. Math. **57** (1977), 45.

[5] H.P. Greenspan, J. Theor. Biol. **65** (1977), 79.

[6] T. Sapir and A. Nir, Physicochem. Hydrodyn. **6** (1985), 803.

[7] J.G. White and G.C. Borisy, J. Theor. Biol. **101** (1983), 289.

[8] J.M. Burgers, *Second Report on Viscosity and Plasticity*, Chap. 3, Kon. Ned. Akad. Wed. Verhand (Eerste Sectie) **16** (1938), 113.

[9] R.G. Cox, J. Fluid Mech. **37** (1969), 601.

[10] J. Opas and M.S. Soltynska, Exp. Cell. Res. **113** (1978), 208.

Stationary Flows in Viscous Fluid Bodies*

Josef Bemelmans

Consider a drop of a viscous, incompressible fluid under the influence of some exterior force density f. A stationary flow inside the fluid body can be described by the Navier-Stokes system

$$-v\Delta v + Dp + v \cdot Dv = f \qquad \text{in } \Omega, \tag{1}$$
$$\text{div } v = 0$$

together with the boundary conditions

$$v \cdot n = 0, \; t_k \cdot T \cdot n = 0 \qquad \text{on } \Sigma, k = 1, 2, \tag{2}$$

$$n \cdot T \cdot n = p_0 \qquad \text{on } \Sigma. \tag{3}$$

As usual, $v = (v^1, v^2, v^3) = v(x)$, $x = (x^1, x^2, x^3)$ denotes the velocity, $p = p(x)$ the pressure, and $v > 0$ is the kinematical viscosity. The unknown domain occupied by the fluid is denoted by Ω, its boundary by Σ; n is the outer normal to Σ, and t_1, t_2 span the tangent plane.

With T being the stress tensor,

$$T_{ij} = -p\delta_{ij} + v(D_i v^j + D_j v^i),$$

$D_i = \partial/\partial x^i$, $D = (D_1, D_2, D_3)$, the dynamical boundary conditions in (2) state that the fluid cannot resist tangential stresses. Equation (3) governs the free boundary.

Σ adjusts itself such that the fluid's normal stress equals the given outer pressure p_0 which is assumed to be constant throughout $\mathbb{R}^3 \backslash \bar{\Omega}$. The volume

---

*In this note we give an outline of results that are contained in a forthcoming paper, *On a free boundary problem for the stationary Navier-Stokes equations.* (Report No. 5, 1986, Institut Mittag-Leffler, Djursholm, Sweden)

$|\Omega|$ of the drop is prescribed, too, and after a suitable normalization we get $|\Omega| = (4/3)\pi$.

The main result is contained in the following theorem; some related results are mentioned at the end of this note.

Theorem. *Let* $f_0(x) = DU(x) = D\int_\Omega g|x - y|^{-1} \, dy$ *be the force of self-attraction. For* $f = f_0 + h$, $h \in C^{s+\mu}$ *with* $h = h(r, x^3) = -h(r, -x^3)$, $r^2 = (x^1)^2 + (x^2)^2$ *and* $\|h\|_{C^{s+\mu}}$ *small enough, there exists a unique solution* $v \in C^{4+\mu}(\bar{\Omega})$, $p \in C^{3+\mu}(\bar{\Omega})$, *and* $\Sigma \in C^{5+\mu}$ *to the free boundary problem* (1)–(3). v *and* p *are small in the sense that*

$$\|v\|_{C^{4+\mu}} + \|p - U\|_{C^{3+\mu}} \leqslant C\|h\|_{C^{s+\mu}}, \tag{4}$$

and Σ *lies in a* $C^{5+\mu}$*-neighborhood of the unit sphere* S, *the* $C^{5+\mu}$*-norm of the distance of* Σ *from* S *can again be estimated by* $C\|h\|_{C^{s+\mu}}$.

The methods to prove this theorem can be explained by comparing them with a device which was used by Solonnikov to solve the time-dependent analogue to (1)–(3). If we denote by $\Omega(t)$ the (unknown) domain occupied by the fluid at time t, Solonnikov [4] used the transformation

$$(x, t) = \left(X + \int_0^t V(X, \tau) \, d\tau, t \right) \tag{5}$$

with $X \in \Omega(0)$, $V(X, \tau)$ being the velocity of the particle initially at X, to map $\bigcup_{0 < t < T} \Omega(t)$ onto the space-time cylinder $\Omega(0) \times (0, T)$, which is now a known domain since the initial value $\Omega(0)$ is given.

As the transformation (5) contains the unknown velocity field, it leads to highly nonlinear equations. However, the kinematical boundary condition for $v \cdot n$ is automatically satisfied, and hence the Navier-Stokes system with four conditions on the free boundary is reduced to a Neumann problem on a given boundary.

This elegant device cannot be used for the stationary problem, because a flow is called stationary if v and p, regarded as functions of the position x rather than of the Lagrange variable X, do not depend on time; as functions of X a stationary velocity field does change in time except for trivial cases.

Our theorem is proved by a perturbation argument; the solution we get for $h \equiv 0$ lies in a neighborhood of the static equilibrium configuration $v \equiv 0$, $p = \int_\Omega g|x - y|^{-1} \, dy$, $\Sigma = S$. We therefore restrict ourselves to free boundaries Σ that can be represented as graphs over the unit sphere, i.e.

$$\Sigma = \{(\xi, \rho): \rho = 1 + \zeta(\xi), \zeta: S \to \mathbb{R}, \xi \in S\}. \tag{6}$$

Let σ be a diffeomorphism of $\bar{\Omega}$ onto the unit ball $\bar{B}$; σ depends on ζ, and if we write the Navier-Stokes equations (1) and the boundary conditions in terms of $y \in \bar{B}$ we obtain a second order differential operator L for the new unknowns $u = u(y)$, $q = q(y)$ that are defined on $\bar{B}$, and the coefficients of L depend on σ. We set

$$\begin{cases} (\xi, \rho) \to \sigma(\xi, \rho) := \left(\xi, \dfrac{\rho}{1 + \zeta(\xi)} \right) \\[3mm] u^i(y) = \left(\det \dfrac{\partial y^i}{\partial x^j} \right)^{-1} \dfrac{\partial y^i}{\partial x^j} v^j(x) \\[3mm] q(y) = p(x) \end{cases} \tag{7}$$

and obtain the transformed equations and boundary conditions

$$Lu^i + \bar{a}_{ij} D_j q + N_i(u, Du) = \tilde{a}_{ij} f^j \qquad \text{in } B \tag{8}$$
$$\operatorname{div} u = 0$$

$$\alpha_i u^i = 0, \; \alpha_{kij} D_i u^j + \beta_{kj} u^j = 0 \qquad \text{on } S \tag{9}$$

$$\alpha_{ij} D_i u^j + \beta_j u^j = 0 \qquad \text{on } S. \tag{10}$$

L is of the form

$$Lu^i = -v D_k(a_{kl} D_l u^i) + b_{ikl} D_k u^l + c_{ij} u^j, \tag{11}$$

N is the transformed nonlinearity

$$N_i(u, Du) = a^{-1} u^j D_j u^i + \tilde{b}_{ikl} u^k u^l. \tag{12}$$

All coefficients depend on σ and its derivatives; the coefficients a_{ij}, α_{ij} etc. on σ and $D\sigma$, b_{ikl}, β_{kj} on σ, $D\sigma$, and $D^2\sigma$, and c_{ij} on σ and its derivatives up to order three.

Schauder estimates for (1), (2) are of the form

$$\|v\|_{C^{2+\mu}} + \|p\|_{C^{1+\mu}} \leqslant C(\|\Sigma\|_{C^{3+\mu}}) \|\varphi\|_{C^{0+\mu}}; \tag{13}$$

that the $C^{3+\mu}$-norm of Σ enters here is due to the fact that the coefficients in the Dirichlet boundary conditions $v \cdot n = 0$ depend on the first derivative of Σ. Solutions to (8)–(10) satisfy the same estimate (13).

We now want to solve (8)–(10) by successive approximations. Let us for the moment assume that (8), (9) can be solved for given σ, that is for a known boundary Σ, and that ζ can be determined from (10) if u and q are known. It is then obvious from (13) that a solution (u, q) of class $C^{2+\mu} \times C^{1+\mu}$ to the Navier-Stokes system can be obtained if σ is of class $C^{3+\mu}$. If we insert this solution (u, q) into (10) to obtain the next approximation to the free boundary Σ, we get a differential equation in ζ with coefficients of class $C^{1+\mu}$. Hence if (10) can be solved at all, its solution is at most $C^{1+\mu}$, too, which means that in every approximation step we encounter a loss of two derivatives. To overcome this difficulty, hard implicit function theorems have been developed. But before we state which version of the Nash–Moser theorem is particularly suited to the problem we consider here, let us turn briefly to equation (10) for the free boundary.

The free boundary ζ is determined by a first order differential operator, and due to the lack of maximum principles for v and p it seems not to be solvable in the general case. Therefore we introduce the force f_0 of self-attraction; as it is a gradient, its potential U can be absorbed into the pressure in (1) and hence (3) becomes

$$\int_\Omega g|x - y|^{-1} dy = -p + v(D_i v^j + D_j v^i)n^i n^j. \tag{14}$$

Note that the unknown ζ appears in the domain of integration Ω. We now regard the left hand side as the main part of equation (14), and according to Lichtenstein [2] the integral $\int_\Omega g|x - y|^{-1} dy$ can be written in the form

$$\psi_0(\xi)\zeta(\xi) + \oint_S \frac{\zeta(\eta)}{d(\xi,\eta)} d\sigma(\eta) + N(\zeta)(\xi) \equiv M(\zeta)(\xi) + N(\zeta)(\xi), \tag{15}$$

where

$$\psi_0(\xi) = \frac{\partial}{\partial n_0} U_0(\xi)$$

$$N(\zeta)(\xi) = R_1(\zeta)(\xi) + R_2(\zeta)(\xi) + R_3(\zeta)(\xi)$$

$$R_1(\zeta)(\xi) = \int_0^{\zeta(\xi)} \left\{ \frac{\partial}{\partial n_0} U(x_0 + \tau n_0) - \frac{\partial}{\partial n_0} U(x_0) \right\} d\tau$$

$$R_2(\zeta)(\xi) = \zeta(\xi) \int_{\mathscr{S}} |x_0 - y|^{-1} dy$$

$$R_3(\zeta)(\xi) = \oint_S \int_0^{\zeta(\xi)} \left\{ \frac{1}{|x_0 - y|} \frac{d\bar{\sigma}_\tau}{d\sigma} - \frac{1}{|x_0 - y|} \right\} d\tau.$$

Here U_0 denotes the volume potential of B, $\mathscr{S} = (\Omega\backslash B) \cup (B\backslash\Omega)$, n_0 is the normal to S, and $d\bar{\sigma}_\tau$ is the surface element $d\sigma_\tau$ of $\Sigma_\tau = \{x = x_0 + \tau n_0 : x_0 \in S\}$ multiplied by $\cos \varphi_\tau$, where φ_τ is the angle between the normals n_0 and n_τ at $x_0 + \tau n_0$; $d(\xi, \eta)$ is the euclidean distance between ξ and $\eta \in S$.

As shown by Lichtenstein, the solvability of the integral equation $M(\zeta) + N(\zeta) = \gamma$ for a given γ is reduced to an investigation of the linear integral operator $M(\zeta)$.* Under the symmetry conditions imposed on h in our theorem, M is invertible and to $v \in C^{2+\mu}$, $p \in C^{1+\mu}$ we obtain a solution $\zeta \in C^{1+\mu}(S)$, the new approximation of the free boundary.

In this way the introduction of f_0 as dominating force leads to an equation for the free boundary which can be handled. But also for physical reasons, f_0 must be regarded as necessary. Self-attraction tends to hold the drop together and therefore balances other forces, like h, that act the opposite way.

There exists a counterexample due to McCready [3] which, although established in quite a different context, supports our interpretation of f_0 as

---

*It is perhaps appropriate to remark that Lichtenstein's equation (15) presents an enormous simplification in the theory of equilibrium figures of rotating liquids. The nonlinearity $N(\zeta)$ is actually of order $o(\zeta)$ and is a really simple expression compared to the involved equations considered by Ljapounoff; already a look at Lichtenstein's lucid presentation of Ljapounoff's results in his monograph [3] is convincing, not to mention Ljapounoff's original papers.

being necessary. McCready shows that it is impossible to give an *a priori* bound for Dirichlet's integral of a solution v to the Navier-Stokes equations if Neumann rather than Dirichlet conditions are given. Hence the basic *a priori* estimate which was first proved by Leray to establish the existence of a global solution (i.e. a solution for arbitrary large forces) fails in this case. In the context of free boundary problems this example indicates that only local solutions exist because there are no longer rigid walls which hold the fluid together regardless of the forces that generate its motion.

We now regard the boundary value problem (8), (9) together with the equation

$$M(\zeta) + N(\zeta) + \frac{1}{g}G(u, q) = 0, \qquad (16)$$

where $G(u, p)$ is the transformation of $-p + v(D_i v^j + D_j v^i)n^i n^j$, as a nonlinear equation

$$F(z) = k,$$

$z = (u, q, \zeta)$, $k = (h, 0, 0)$ which is to be solved in a neighborhood of $z_0 = (0, U, 0)$, the solution to

$$F(z_0) = 0.$$

If we define $\Sigma_1 + \Sigma_2$ where $\Sigma_i = \{(\xi, \rho): \rho = 1 + \zeta_i(\xi)\}$ to be the surface $\{(\xi, \rho): \rho = 1 + \zeta_1(\xi) + \zeta_2(\xi)\}$ the set of functions on which F acts, bears an affine structure.

F then is a mapping from $\mathscr{X}_0 = C^{2+\mu}(\bar{\Omega}, \mathbb{R}^3) \times C^{1+\mu}(\bar{\Omega}, \mathbb{R}) \times C^{3+\mu}(S, \mathbb{R})$ into $\mathscr{Y}_0 = C^{0+\mu}(\bar{\Omega}, \mathbb{R}^5)$, as well as from $\mathscr{X}_t = C^{2+\mu+t} \times C^{1+\mu+t} \times C^{3+\mu+t}$ into $\mathscr{Y}_t = C^{0+\mu+t}$, $t > 0$. Its derivative can be defined then:

$$DF(z)\tilde{z} = \lim_{\varepsilon \to 0} \frac{1}{\varepsilon}\{F(z + \varepsilon\tilde{z}) - F(z)\}.$$

The application of hard implicit function theorems rests on $DF(z)$ to be invertible not only at z_0 but in a full neighborhood of z_0; the loss of derivatives one encounters then is compensated by applying suitable smoothing operators. If Moser's theorem is applied, the error introduced by the smoothing operators is compensated by the fact that one uses a rapidly convergent iteration scheme.

If one calculates $DF(z)\tilde{z}$ on the basis of (8), (9), and (16), one arrives at a set of linearized equations; the formal but tedious manipulations are given in our paper already referred to. Of interest is only the following property: the linearization of (8) consists of a second order differential operator in $\tilde{u}^i$ and $\tilde{q}$ (basically the Stokes equations) together with a third order operator in $\tilde{\sigma}$ which is the transformation (7) belonging to $\tilde{\zeta}$:

$$L(u, q, \zeta)u^i + \bar{a}_{ij}(\zeta)D_j\tilde{q} + l_{ijk\gamma}(u, q, \zeta)D^\gamma\sigma^{jk} = \tilde{\alpha}_{ij}h^j, \qquad (17)$$

where γ is a multi-index with $|\gamma| \leq 3$. Obviously the expression in $\tilde{\sigma}$ can never be regarded as a small perturbation, therefore (17) is not solvable.

For this reason we use a variant of the implicit function theorem which is due to Zehnder [5]. It is especially suited to the above problem because it allows one to work with a weaker form of DF. Instead of requiring the existence of the inverse of DF, it suffices to construct an operator T, such that

$$|(DF(z) \circ T(z, z') - 1|)(\tilde{z})| \leqslant C |z| |F(z) - k|. \tag{18}$$

T serves as an approximate inverse in the sense that it approaches $(DF)^{-1}$ as the sequence of successive approximations tends to the solution of $F(z) = k$. It is essential that (18) still leads to a rapidly convergent method because of the factor $|F(z) - k|$. In the n-th approximation we have to solve (17) with $(u_{n-1}, q_{n-1}, \zeta_{n-1}) = z_{n-1}$ to obtain $\tilde{z}_n$; in the last expression we replace the unknown $\tilde{\sigma}_n$ by the quantity $\tilde{\sigma}_{n-1}$ that was established in the previous step. In this way (17) becomes an equation of Stokes type in the unknown $\tilde{u}^i$ and $\tilde{q}$ alone. Existence and regularity for the solution to this equation now presents no difficulties.

The same type of problem occurs in (16); it can be handled in precisely the same way. In solving this equation the symmetry assumption on h that was stated in the theorem becomes necessary.

Infinitesimal translations of S in the direction of the coordinate axes are always solutions to the linear part of (16); hence the right hand side must be perpendicular to them with respect to the L_2-scalar product. This is established, e.g. by the symmetry of h. Physically it means that the forces that generate a flow inside the fluid body must be balanced in the sense that the barycenter of Ω is kept fixed. Once this setup is established, the proof follows along the lines of Zehnder's theorem.

Extension of the results: Thus far we never used the fact that the solution z_0 consists of a ball with no flow inside. Lichtenstein's equation applies to all equilibrium figures of rotating liquids, and we can use instead of z_0 from above any such figure provided it is isolated in the following sense. For certain values of the angular velocity, branching occurs; from the family of McLaurin ellipsoids, for instance, the ellipsoids of Jacobi branch off. This phenomenon is reflected in the fact that the linearization of (16) admits non-trivial solutions.

Our analysis does not cover this phenomenon; hence we require that the initial configuration z_0 is separated from critical values of this type.

References

[1] L. Lichtenstein, *Gleichgewichtsfiguren rotierender Flüssigkeiten*, Berlin (1933).

[2] L. Lichtenstein, *Zur Theorie der Gleichgewichtsfiguren rotierender Flüssigkeiten*, Math. Z. **39** (1935), 639–648.

[3] T.A. McCready, *The interior Neumann problem for stationary solutions of the Navier-Stokes equations*, Dissertation, Stanford Univ. (1968).

[4] V.A. Solonnikov, *Solvability of a problem on the motion of a viscous, incompressible fluid bounded by a free surface*, Math. U.S.S.R. Izvestija **11** (1977), 1323–1358.

[5] E. Zehnder, *Generalized Implicit Function Theorems with Applications to Some Small Divisor Problems*, I, Comm. Pure Appl. Math. **28** (1975), 91–140.

Large Time Behavior for the Solution of the Non-Steady Dam Problem

Dietmar Kröner

1. Introduction and Main Results

In this paper we consider two water reservoirs which are separated by a dam D consisting of an isotropic, homogeneous, porous material. The levels of the reservoirs may be different and time dependent and they are supposed to tend to fixed levels if t tends to infinity. We start with nonstationary initial conditions and we are interested in the asymptotic behavior of the pressure distribution $u(t, z)$ of the water in the dam if t tends to infinity.

Let us start with the assumptions concerning ∂D. The boundary is supposed to satisfy:

$$\partial D = \bar{\Gamma}_0 \cup \bar{\Gamma}_1, \Gamma_0 \cap \Gamma_1 = \varnothing; \Gamma_0, \Gamma_1 \text{ are}$$

relatively open; ∂D is Lipschitz continuous

and Γ_0, Γ_1 are graphs of functions $\psi_0, \psi_1 \in C^2([a, b])$

such that $D = \{(x, y) | x \in]a, b[, \psi_1(x) < y < \psi_0(x)\}$.

$e \cdot v < 0$ on Γ_1 where e is the $\hspace{3em}$ (1.1)

vertical unit vector $(0, 1)$ and v is the

outward normal to Γ_1. Let $P := (x_0, y_0)$

denote the top of the dam. Then $\psi_0'(x) > 0$

for $a < x < x_0$ and $\psi_0'(x) < 0$ for $x_0 < x < b$.

The last conditions ensure that the number of reservoirs remains constant.

On Γ_1 the dam is assumed to be impervious. We split up Γ_0 into $\Gamma_{01} := \{(x, y) \in \Gamma_0 | a \leqslant x \leqslant x_0\}$, $\Gamma_{02} := \{(x, y) \in \Gamma_0 | x_0 \leqslant x \leqslant b\}$ and assume:

$$u(t, x, y) = (y_i(t) - y)^+ \quad \text{in } \Sigma_{0i} := {]}0, \infty{[} \times \Gamma_{0i},$$

$$0 \leqslant y_i(t) < y_0 \quad \text{for } t \in {]}0, \infty{[} \tag{1.2}$$

$$y_i(t) \to Y_i \quad \text{if } t \to \infty \; i = 1, 2.$$

Here $y_i(t)$ measures the water levels of the reservoirs $i = 1, 2$ at time t, and they are supposed to stay below the maximal height y_0 of the dam. Y_i, $i = 1, 2$ defines the height of the ith-reservoir in the stationary situation.

Initially we prescribe the wet part D_0 of the dam. Let χ_0 denote the characteristic function of $D_0 \subset D$. We assume that the following compatibility condition is fulfilled:

$$\chi_0|_{\Gamma_0} = \text{sign}(u|_{\{0\} \times \Gamma_0}). \tag{1.3}$$

For the weak formulation of the problem we need some further technical assumptions. Let us assume that there exists an extension g of the boundary values on Γ_0 such that

$$g \in C^{0,1}(\bar{D}_\infty) \cap L^\infty(D_\infty), \quad g \geqslant 0 \quad \text{in } D_\infty;$$

$$g(t, x, y) = (y_i(t) - y)^+ \quad \text{on } \Sigma_{0i}, \; i = 1, 2. \tag{1.4}$$

Since we shall look for the asymptotic behavior for $t \to \infty$, we suppose that there exists a function G such that

$$G \in C^{0,1}(D),$$

$$g(t, \cdot) \to G \quad \text{a.e. in } D \text{ if } t \to \infty, \tag{1.5}$$

$$G(x, y) = (Y_i - y)^+ \quad \text{on } \Gamma_{0i}, \; i = 1, 2.$$

The space of testfunctions is

$$V = \{\omega \in H^{1,2}(D) | \omega = 0 \text{ on } \Gamma_0\}.$$

Then we shall investigate the following weak formulation of the described problem.

1.1 NON-STATIONARY PROBLEM. The data D, g, χ_0 are supposed to satisfy (1.1), (1.3), (1.4). Then find a pair of functions $\{u, \gamma\}$ such that we have

$$u \in g + L^2_{\text{loc}}(0, \infty; V), \quad \gamma \in L^\infty(D_\infty), \quad \partial_t \gamma \in L^2_{\text{loc}}(0, \infty; V^*);$$

$$u \geqslant 0, \quad 0 \leqslant \gamma \leqslant 1, \quad u(1 - \gamma) = 0 \quad \text{a.e. in } D_\infty;$$

$$\int_{D_\infty} (\gamma(\partial_y v - \partial_t v) + \nabla u \nabla v) \leqslant 0$$

for all $v \in \mathring{H}^1(0, \infty; H^1(D))$, $v \geqslant 0$ on $\Gamma_0 \cap \{g = 0\}$, $v = 0$ on $\Gamma_0 \cap \{g > 0\}$,
$$\tag{1.6}$$

$$\int_0^T \int_D (\gamma - \gamma_0) \partial_t \zeta = \int_0^T \langle \partial_t \gamma, \zeta \rangle$$

for all $\zeta \in L^2(0, T; V) \cap H^1(0, T, L^\infty(D))$, $\zeta(T) = 0$ for all $T \in \mathbb{R}^+$.

In this paper we intend to show, that the solution of Problem 1.1 which we get by regularization converges strongly in $L^p(D)$, $1 \leqslant p < \infty$ to the solution u_∞ of the stationary problem which is defined in Problem 1.2.

1.2 STATIONARY PROBLEM. For given data D and G satisfying (1.1), (1.5) find $\{u_\infty, \gamma_\infty\}$ such that we have

$$u_\infty \in G + V, \quad \gamma_\infty \in L^\infty(D);$$

$$u_\infty \geqslant 0, \quad 0 \leqslant \gamma_\infty \leqslant 1, \quad u_\infty(1 - \gamma_\infty) = 0 \qquad \text{a.e. in } D;$$

$$\int_D (\nabla u_\infty + e\gamma_\infty)\nabla v \leqslant 0 \qquad \begin{array}{l} \text{for all } v \in H^1(D), v \geqslant 0 \text{ on } \Gamma_0 \cap \{G = 0\}, \\ v = 0 \text{ on } \Gamma_0 \cap \{G > 0\}. \end{array}$$

1.3 Remark. We have to assume that

$$\text{Problem 1.2 has at most one solution.} \qquad (1.7)$$

Conditions for the data of Problem 1.2 under which (1.7) is true can be found in [AG], [BR], and [CC].

Let us continue with some known results in this field. For Problem 1.1, Gilardi ([GI], Theorem 4.1) has proved the existence of at least one bounded solution. In the case where we have $\theta(u)$ instead of γ in (1.6) and θ is Lipschitz continuous the global behavior was studied in [KR]. There it was even shown that $u(t) \to u_\infty$ strongly in $H^1(D)$ for $t \to \infty$ and the rate of convergence could be estimated. Using the Baiocchi transformation, Rodrigues [R] has derived the convergence for $t \to \infty$ for the time dependent rectangular dam problem. In a recent paper, Friedman and DiBenedetto [FB] consider a rectangular dam which separates compressible fluids. If the levels of the reservoirs are moving periodically in time they show that there exist uniquely determined initial values for which the corresponding solution is periodical in time with the same period. For investigations concerning the asymptotic behavior of solutions of Stefan-type problems, see for example [KE].

The existence of the solution of the stationary problem we get from the following.

1.4 Theorem (existence for the stationary problem). *There exists a solution of Problem* 1.2.

PROOF. See Alt [A3] [BR]. □

Remark. In this paper we do not assume that Problem 1.2 has a solution. In §4 we shall show independently of Theorem 1.4 that there exists a solution of Problems 1.2.

Now let us describe the regularization which will give us a solution of the non-stationary problem 1.1. We define

$$b_\varepsilon(t) := \begin{cases} 1 & \text{if } t \geq \varepsilon \\ \dfrac{t}{\varepsilon} & \text{if } 0 < t < \varepsilon. \\ 0 & \text{if } t \leq 0 \end{cases} \qquad (1.8)$$

For approximating a solution of Problem 1.1 we consider the following regular problem.

1.5 REGULAR PROBLEM. The data D, g, χ_0, b_ε are supposed to satisfy (1.1), (1.3), (1.4), (1.8) and assume

$$u_0 \in g(0, \cdot) + V \cap L^\infty(D), \quad u_0 > 0 \text{ in } D_0, \quad u_0 = 0 \text{ in } D \backslash D_0. \qquad (1.9)$$

Then find a function u_ε such that we have

$$u_\varepsilon \in g + L^\infty_{\text{loc}}(0, \infty; V), \quad \partial_t b_\varepsilon(u_\varepsilon) \in L^2_{\text{loc}}(0, \infty; L^2(D));$$

$$u_\varepsilon \geq 0 \quad \text{in } D_\infty;$$

$$\int_D \partial_t b_\varepsilon(u_\varepsilon) v + \int_D (\nabla u_\varepsilon + e b_\varepsilon(u_\varepsilon)) \nabla v = 0 \qquad \text{for all } v \in V; \qquad (1.10)$$

$$b_\varepsilon(u_\varepsilon(0, \cdot)) = b_\varepsilon(u_0) \qquad \text{on } \{0\} \times D.$$

1.6 Theorem. *There exists for any $\varepsilon > 0$ one and only one solution u_ε of Problem 1.5.*

PROOF. See Alt and Luckhaus [AL] 2.2 and 2.3. □

If ε tends to zero, the solutions u_ε of Problem 1.5 converge to a solution of Problem 1.1. This is the assertion of the following theorem.

1.7 Theorem. *Suppose (1.1), (1.3), (1.4), (1.5), (1.8), and (1.9). Then there exists a subsequence u_ε of solutions of the regular problem 1.5 such that u_ε converges weakly in $L^2_{\text{loc}}(0, \infty; V)$ to u and $b_\varepsilon(u_\varepsilon)$ converges weakly in $L^p_{\text{loc}}(D_\infty)$ to γ, $1 \leq p \leq \infty$, where $\{u, \gamma\}$ is a solution of the non-stationary problem 1.1.*

PROOF. See §4. □

Now we can formulate the main result of this paper.

1.8 Theorem. *Suppose (1.1), (1.3), (1.4), (1.5), (1.7), (1.8), and (1.9). Let $\{u, \gamma\}$ be the solution of Problem 1.1 which we get in Theorem 1.7. Let $\{u_\infty, \gamma_\infty\}$ be the solution of the stationary problem 1.2. Then we have*

$$u(t) \to u_\infty, \gamma(t) \to \gamma_\infty \qquad \text{for } t \to \infty$$

strongly in $L^p(D)$ for all $1 \leq p < \infty$.

We shall give the proof in the remaining part of this paper (see 4.4).

2. Sub- and Supersolutions

The main idea for proving Theorem 1.8 is to construct sub- and supersolutions u^+, u^- for u, which are decreasing and increasing in t, respectively, if t tends to infinity. Therefore we have to choose suitable boundary and initial values for u_ε^+ and u_ε^-. Since we must estimate the measure of the sets $\{z \in D \mid 0 < G(z) < \varepsilon\}$ in terms of ε, it turns out to be successful to replace G by a harmonic function with the same boundary values on Γ_0 as G. Let $z_i = (x_i, Y_i) \in \Gamma_0$, $i = 1, 2$ denote the points where the surfaces of the water reservoirs touch the dam in the steady-state and $S_1 := \{z = (x, y) \in \Gamma_0 \mid x_1 \leqslant x \leqslant x_2\}$.

2.1 Lemma. *Assume* (1.1), (1.5). *Then there exists a solution* $H \in G + V$ *of*

$$\int_D (\nabla H + e)\nabla v = 0 \qquad \text{for all } v \in V. \tag{2.1}$$

For H we have the following properties:

$$H \in L^\infty(D) \qquad \text{and } H > 0 \text{ in } D. \tag{2.2}$$

There exist ρ_0, ε_0 such that for $i = 1, 2$:

$$|\nabla H(z)| = \mathcal{O}(|\log|z - z_i||) \qquad \text{for all } z \in B_{\rho_0}(z_i) \cap D;$$

$$\partial_y H \leqslant -k < 0 \qquad \text{on } D^1_{\rho_0} := \{z \in D \mid \text{dist}(z, S_1) \leqslant \rho_0\} \text{ for some } k > 0;$$

$$\text{meas}(\{z \in D \mid a \leqslant H(z) \leqslant \varepsilon + a\} \cap D_{\rho_0}) \leqslant k_1 \varepsilon \qquad \text{for } a \in \mathbb{R}, 0 < \varepsilon \leqslant \varepsilon_0,$$

$$\text{where } k_1 \text{ is independent of } a \text{ and } D_{\rho_0} := \{z \in D \mid \text{dist}(\Gamma_0, z) \leqslant \rho_0\}; \tag{2.3}$$

$$\{z \in D \mid 0 \leqslant H \leqslant 2\varepsilon\} \subset D^1_{\rho_0};$$

$$\int_{\{a \leqslant H \leqslant \varepsilon + a\} \cap D^1_{\rho_0}} |\nabla H|^2 = \mathcal{O}(\varepsilon), \qquad 0 < \varepsilon \leqslant \varepsilon_0.$$

PROOF. See [K], Lemma 3.1. □

As boundary values for the sub- and supersolution we choose

$$F^\pm(t, z) := H(z) \pm \varphi^\pm(t) \qquad (t, z) \in [0, \infty[\times D,$$

$$u_0^+(z) := H(z) + \varphi^+(0) \qquad z \in D, \tag{2.4}$$

$$u_0^- \text{ solves: } \int_D \nabla u_0^- \nabla v = 0 \qquad \text{for all } v \in V, u_0^- \in F^-(0, \cdot) + V,$$

such that φ satisfies

$$H(z) - \varphi^-(t) \leqslant g(t, z) \leqslant H(z) + \varphi^+(t) \qquad \text{on }]0, \infty[\times \Gamma_0,$$

$$H(z) - \varphi^-(0) \leqslant u_0(z) \leqslant H(z) + \varphi^+(0) \qquad \text{on } D \tag{2.5}$$

and

$\|\varphi^{\pm}\|_{C^2([0,\infty[)}$ bounded; $\varphi^{\pm} > 0$, $(\varphi^{\pm})' \leqslant 0$; $\qquad \varphi^{\pm}(t) \to 0$ if $t \to \infty$;

$$(\varphi^+)' \geqslant -k \qquad \text{on } [0,\infty[\text{ where } k \text{ is given in (2.3);}$$

$$(\varphi^-)' \leqslant -K, \quad K := \sup\{\partial_y H(z) | z \in D \setminus D^1_{\rho_0}\} \qquad \text{for } 0 < t < t_0 \tag{2.6}$$

$$\text{and } \varphi^-(t) \leqslant \varepsilon_0 \text{ for } t < t_0.$$

This can be obtained, for example, if we choose $\varphi^+(0)$ and $\varphi^-(0)$ large enough. Then the sub- (super-) solutions $u^{\pm}_{\varepsilon}$ are defined as follows:

$$u^{\pm}_{\varepsilon} \in F^{\pm} + L^{\infty}_{\text{loc}}(0,\infty; V), \quad \partial_t b_{\varepsilon}(u^{\pm}_{\varepsilon}) \in L^2_{\text{loc}}(0,\infty; L^2(D));$$

$$\int_D \partial_t b_{\varepsilon}(u^{\pm}_{\varepsilon})v + \int_D (\nabla u^{\pm}_{\varepsilon} + e b_{\varepsilon}(u^{\pm}_{\varepsilon}))\nabla v = 0 \qquad \text{for all } v \in V; \tag{2.7}$$

$$b_{\varepsilon}(u^{\pm}_{\varepsilon}(0,\cdot)) = b_{\varepsilon}(u^{\pm}_0) \qquad \text{on } D.$$

Remark. For existence and uniqueness we use again [AL], Theorem 2.3 and 2.4.

2.2 Theorem. u^+_{ε} and u^-_{ε} *are monotone decreasing and increasing in t, respectively. Furthermore,*

$$u^-_{\varepsilon} \leqslant u_{\varepsilon} \leqslant u^+_{\varepsilon}, \quad u^-_0 \leqslant u^-_{\varepsilon} \leqslant F^-, \quad 0 \leqslant u^+_{\varepsilon} \leqslant F^+, \qquad \text{a.e. in } D_{\infty}.$$

PROOF. We confine ourselves to the proof of $u^-_{\varepsilon} \leqslant F^-$. For more details see [K]. For the boundary and initial values the inequality is true by definition of F^-. For F^- we get for $v \in V$, $v \geqslant 0$

$$\int_D \partial_t b_{\varepsilon}(F^-)v + \int_D (\nabla F^- + e b_{\varepsilon}(F^-))\nabla v$$

$$\geqslant \frac{1}{\varepsilon} \int_{\{0 \leqslant F^- \leqslant \varepsilon\} \cap D^1_{\rho_0}} v((-\varphi^-)' - \partial_y H) + \frac{1}{\varepsilon} \int_{\{0 \leqslant F^- \leqslant \varepsilon\} \setminus D^1_{\rho_0}} v((-\varphi^-)' - \partial_y H).$$

The first integral is nonnegative since $(\varphi^-)' < 0, v \geqslant 0$ and $\partial_y H \leqslant 0$ in $D^1_{\rho_0}$. For the second integral we distinguish between the cases $t < t_0$ and $t \geqslant t_0$, where t_0 is defined in (2.6). If $t \leqslant t_0$ we have $(\varphi^-)' \leqslant -K$ and therefore

$$\int_{\{0 \leqslant F^- \leqslant \varepsilon\} \setminus D^1_{\rho_0}} v(-(\varphi^-)' - \partial_y H) \geqslant \int_{\{0 \leqslant F^- \leqslant \varepsilon\} \setminus D^1_{\rho_0}} v(K - \partial_y H) \geqslant 0.$$

If $t > t_0$ we obtain

$$\{(t,z)|0 \leqslant F^-(t,z) \leqslant \varepsilon\} \setminus D^1_{\rho_0} = \{0 \leqslant H(z) - \varphi^-(t) \leqslant \varepsilon\} \setminus D^1_{\rho_0}$$

$$= \left\{0 \leqslant \frac{H}{2} + \left(\frac{H}{2} - \varphi^-(t)\right) \leqslant \varepsilon\right\} \setminus D^1_{\rho_0}$$

$$\subset \left\{0 \leqslant \frac{H}{2} \leqslant \varepsilon\right\} \setminus D^1_{\rho_0} = \varnothing$$

because of (2.3), (2.6) and because of $H \geqslant 2\varepsilon_0$ in $D \setminus D^1_{\rho_0}$. Then applying the comparison theorem ([AL], Theorem 2.2) we obtain $u^-_\varepsilon \leqslant F^-$. $\square$

3. $L^\infty(H^1)$-Estimates for Sub- and Supersolutions

Let us use the following notations:

$$\partial^h_t w(t, z) := \frac{1}{h}(w(t + h, z) - w(t, z)),$$

$$w_h(t, z) := \frac{1}{h} \int_t^{t+h} w(s, z)\,ds,$$

$$B^h_\varepsilon(w, s, z) := \frac{b_\varepsilon(w(s + h, z)) - b_\varepsilon(w(s, z))}{w(s + h, z) - w(s, z)}.$$

3.1 Theorem. *There exist constants* $\varepsilon_0, h_0, t_1, C_0 > 0$ *such that we have for all* $0 < \varepsilon \leqslant \varepsilon_0, 0 < |h| \leqslant h_0$ *and* $t \geqslant t_1$:

$$\int_t^{t+1} \int_D B^h_\varepsilon(w, s, z)\partial^h_t w(s, z)^2\,dz\,ds + \int_D |\nabla w(t, z)|^2\,dz \leqslant C_0$$

for $w = u^+_\varepsilon, u^-_\varepsilon$.

Remark. The first term on the left side of the inequality in Theorem 3.1 can be estimated by C_0/h for all $t \in \mathbb{R}^+$.

PROOF. Using $\partial^h_t w - \partial^h_t F$ as testfunction in (2.7) we obtain

$$A_{\varepsilon h}(t + 1) \leqslant C + \int_t^{t+1} A_{\varepsilon h}(s)\,ds + A_{\varepsilon h}(t) - A_{\varepsilon h}(t + 1) + 4(w^t_\varepsilon(h) + M^t_\varepsilon(h))$$

where

$$A_{\varepsilon h}(t) := \int_D [\nabla w_h(t) + e(b_\varepsilon(w(t)))_h]^2\,dz,$$

$$w^t_\varepsilon(h) := \int_t^{t+1} \int_D |\nabla w_h(s) - \nabla w(s)|^2 B^h_\varepsilon(w, s)\,dz\,ds,$$

$$M^t_\varepsilon(h) := \int_t^{t+1} \int_D |\nabla w_h(s)|^2 B^h_\varepsilon(w, s)\,dz\,ds,$$

and for fixed ε and t we obtain if h tends to zero

$$w^t_\varepsilon(h) \to 0,$$

$$M^t_\varepsilon(h) \to \int_t^{t+1} \int_D |\nabla w(s)|^2 b'_\varepsilon(w(s))\,dz\,ds.$$

Now in order to estimate $M^t_\varepsilon(h)$ we need the following lemma.

3.2 Lemma. *Under the assumptions of Theorem 3.1 there are constants t_1, $C_1 > 0$ such that we have for all $t \geqslant t_1$ and $0 < \varepsilon \leqslant \varepsilon_0$:*

$$\int_D b_\varepsilon(w(s))^2 \Big|_{s=t}^{s=t+1} + \int_t^{t+1} \int_D b_\varepsilon'(w(s)) |\nabla w(s)|^2 \, dz \, ds \leqslant C_1,$$

where $w = u_\varepsilon^\pm$. Therefore we get

$$\limsup_{h \to 0} 2A_{\varepsilon h}(t+1) \leqslant \limsup_h \int_t^{t+1} A_{\varepsilon h}(s) \, ds + \limsup_h A_{\varepsilon h}(t) + C$$

$$\leqslant C + \limsup_h A_{\varepsilon h}(t). \tag{3.1}$$

The second inequality holds since the integral is bounded. This can be seen if we test (2.7) with $w - F$ ([K], Lemma 5.3). Then (3.1) implies (see Lemma 3 in [KR]) $\limsup_h A_{\varepsilon h}(t+1) \leqslant C$ where C is independent of ε and t and therefore the statement of Theorem 3.1.

3.3 PROOF OF LEMMA 3.2. Using $b_\varepsilon(w) - b_\varepsilon(F)$, $w = u_\varepsilon^\pm$, as testfunction in (2.7) and integrating over $]t, t+1[$ we obtain

$$\frac{1}{2} \int_D b_\varepsilon(w)^2 \Big|_t^{t+1} + \int_t^{t+1} \int_D b_\varepsilon'(w) |\nabla w|^2$$

$$\leqslant \int_D b_\varepsilon(w) b_\varepsilon(F)|_t^{t+1} - \int_t^{t+1} \int_D b_\varepsilon(w) b_\varepsilon'(F)\varphi' + \int_t^{t+1} \int_D \nabla w b_\varepsilon'(F) \nabla F$$

$$+ \int_t^{t+1} \int_D b_\varepsilon(w)(\partial_y b_\varepsilon(w) - \partial_y b_\varepsilon(F)), \qquad \varphi = \pm \varphi^\pm. \tag{3.2}$$

Let us demonstrate the principal ideas estimating the third integral on the right side. For more details see [K], 5.4. The third integral on the right side can be estimated as follows:

$$\left| \int_t^{t+1} \int_D \nabla w b_\varepsilon'(F) \nabla H \right| \leqslant \frac{\delta}{2} \int_t^{t+1} \int_D |\nabla w|^2 b_\varepsilon'(F) + \frac{1}{2\delta} \int_t^{t+1} \int_D |\nabla H|^2 b_\varepsilon'(F)$$

$$= \frac{\delta}{2\varepsilon} \int_t^{t+1} \int_{\{0 \leqslant F(s) \leqslant \varepsilon\}} |\nabla w|^2 \, dz \, ds$$

$$+ \frac{1}{2\delta\varepsilon} \int_t^{t+1} \int_{\{0 \leqslant F(s) \leqslant \varepsilon\}} |\nabla H|^2 \, dz \, ds.$$

Let us confine ourselves to the first term. Theorem 2.2 implies

$$\left. \begin{array}{l} u_\varepsilon^- \leqslant u_\varepsilon \leqslant u_\varepsilon^+ \\ u_\varepsilon^- \leqslant F^-, u_\varepsilon^+ \leqslant F^+ \end{array} \right\} \quad \text{a.e. in } D_\infty.$$

We have to distinguish between the cases where $w = u_\varepsilon^+$, $w = u_\varepsilon^-$.

$$w = u_\varepsilon^+ : \quad \{0 \leqslant F^+(s) \leqslant \varepsilon\} = \{z | 0 \leqslant F^+(s,z) \leqslant \varepsilon\}$$
$$\subset \{z | 0 \leqslant u_\varepsilon^+(s,z) \leqslant \varepsilon\} \ (u_\varepsilon^+ \geqslant 0!)$$
$$w = u_\varepsilon^- : \quad \{0 \leqslant F^-(s) \leqslant \varepsilon\} = \{z | 0 \leqslant F^-(s,z) \leqslant \varepsilon\}$$
$$\subset \{z | 0 \leqslant (u_\varepsilon^-(s,z))^+ \leqslant \varepsilon\}$$

where $(a)^+ := \max(a, 0)$.

Hence for $w = u_\varepsilon^-$:

$$\frac{1}{\varepsilon} \int_t^{t+1} \int_{\{0 \leqslant F^-(s) \leqslant \varepsilon\}} |\nabla u_\varepsilon^-|^2 \leqslant \frac{1}{\varepsilon} \int_t^{t+1} \int_{\{0 \leqslant (u_\varepsilon^-(s))^+ \leqslant \varepsilon\}} |\nabla u_\varepsilon^-|^2$$
$$= \int_t^{t+1} \int_D b_\varepsilon'((u_\varepsilon^-)^+)|\nabla u_\varepsilon^-|^2 = \int_t^{t+1} \int_D b_\varepsilon'(u_\varepsilon^-)|\nabla u_\varepsilon^-|^2.$$

The corresponding terms for u_ε^+ can be treated similarly, and in any case, the integrals appear on the left side in (3.2).   □

4. Convergence

In this section we shall study first the convergence of u_ε if ε tends to zero and then if t tends to infinity. The results are collected in several lemmas. For more details of the proofs see [K].

4.1 Lemma. *Under the assumption of Theorem 3.1 there exist u, $u^\pm \in L_{loc}^2(0, \infty; H^1(D))$ such that we have for a suitable subsequence $\varepsilon \to 0$,*

$$u_\varepsilon^\pm \to u^\pm \quad \begin{cases} \text{weakly in } L_{loc}^2(0, \infty; H^{1,2}(D)) \text{ and} \\ \text{weakly star in } L_{loc}^\infty(t_1, \infty; V); \end{cases} \tag{4.1}$$

$$u_\varepsilon \to u \quad \text{weakly in } L_{loc}^2(0, \infty; H^{1,2}(D)).$$

Furthermore

$$\int_D |\nabla u^\pm(t)|^2 \leqslant \text{const}$$

uniformly for all $t \geqslant t_1$. $u^\pm$ are monotone decreasing and increasing in t, respectively, and

$$u^- \leqslant u \leqslant u^+ \quad \text{a.e. in } D_\infty.$$

Essentially this follows from Theorems 2.2 and 3.1.

4.2 Lemma. *Under the assumptions of Theorem 3.1 there exist γ, $\gamma^\pm \in L_{loc}^p(D_\infty)$ such that we have for a suitable subsequence:*

$$\begin{rcases} b_\varepsilon(u_\varepsilon) \to \gamma \\ b_\varepsilon(u_\varepsilon^\pm) \to \gamma^\pm \end{rcases} \quad \begin{array}{l} \text{weakly in } L_{loc}^p(D_\infty), \\ p > 1. \end{array}$$

$\gamma^{\pm}$ *are monotone decreasing and increasing in t, respectively, and*

$$\gamma^- \leqslant \gamma \leqslant \gamma^+ \qquad a.e.\ in\ D_\infty.$$

This is obvious since $0 \leqslant b_\varepsilon \leqslant 1$, b_ε is monotone and because of Theorem 2.2.

4.3 Lemma. *Under the assumptions of Theorem 3.1 we have*

$$u \in g + V,\ u^{\pm} \in F^{\pm} + V,\ \partial_t \gamma,\ \partial_t \gamma^{\pm} \in L^2_{\mathrm{loc}}(0, \infty; V^*),$$

$$\int_0^\infty \int_D \gamma^{\pm}(\partial_y - \partial_t)v + \int_0^\infty \int_D \nabla u^{\pm} \nabla v = 0 \qquad for\ all\ v \in \mathring{H}^1(0, \infty; V).$$

$$\int_0^\infty \int_D \left(\gamma(\partial_y - \partial_t)v + \int_0^\infty \int_D \nabla u \nabla v \right) \leqslant 0 \quad \begin{matrix} for\ all\ v \in \mathring{H}^1(0, \infty; H^1(D)), \\ v \geqslant 0\ on\ \Gamma_0 \cap \{g = 0\}, \\ v = 0\ on\ \Gamma_0 \cap \{g > 0\}. \end{matrix} \qquad (4.2)$$

Furthermore, if b_0 denotes the pointwise limit of b_ε

$$\gamma(0, \cdot) = b_0(u_0)$$

$$\gamma^{\pm}(0, \cdot) = b_0(u_0^{\pm}(0)) = \begin{cases} 1 & if\ \text{``}+\text{''} \\ 0 & if\ \text{``}-\text{''} \end{cases}$$

(in the weak sense, see (1.6)) and

$$u \geqslant 0,\quad 0 \leqslant \gamma \leqslant 1,\quad u(1 - \gamma) = 0 \qquad a.e.\ in\ D_\infty;$$

$$0 \leqslant \gamma^{\pm} \leqslant 1,\quad \begin{cases} u^{\pm} > 0 \Rightarrow \gamma^{\pm} = 1 \\ u^{\pm} < 0 \Rightarrow \gamma^{\pm} = 0 \end{cases} \qquad a.e.\ in\ D_\infty. \qquad (4.3)$$

The most important step in the proof of this lemma is to show that $\gamma \in \lim b_\varepsilon(u)$. For this we apply Lemma 4.3 [AL].

4.4 PROOF OF THEOREM 1.8. Since $\{u^-, \gamma^-\}$, $\{u^+, \gamma^+\}$ are bounded and monotone increasing and decreasing functions in t, respectively (see Lemmas 4.1 and 4.2), we can define the following pointwise limits:

$$u_\infty^{\pm}(z) := \lim_{t \to \infty} u^{\pm}(t, z),$$
$$\gamma_\infty^{\pm}(z) := \lim_{t \to \infty} \gamma^{\pm}(t, z) \qquad for\ a.a.\ z \in D. \qquad (4.4)$$

The $L^\infty(H^1)$ estimates of $u^{\pm}$ in Lemma 4.1 imply that the convergence

$$u^{\pm}(t, \cdot) \to u_\infty^{\pm}, \qquad (4.5)$$

holds weakly in $H^{1,2}(D)$ (for subsequences) and by Lebesgue's convergence theorem strongly in $L^p(D)$ ($1 \leqslant p < \infty$). Moreover

$$u_\infty^{\pm} \in H + V.$$

Now consider the weak equation for $u^{\pm}$ in (4.2). From (4.2) we obtain for any

$t \in \mathbb{R}^+$ for all $v \in \mathring{H}^1(]t, t + 1[, V)$

$$\int_t^{t+1} \int_D \gamma^\pm (\partial_y - \partial_t) v \, dz \, d\tau + \int_t^{t+1} \int_D \nabla u^\pm \nabla v \, dz \, d\tau = 0. \qquad (4.6)$$

Let $\varphi \in C_0^\infty(]0, 1[)$ such that $\int_0^1 \varphi(s) \, ds \neq 0$, $\psi \in V$ and take $v(\tau, z) := \varphi(\tau - t)\psi(z)$, $t \in \mathbb{R}^+$ as testfunction in (4.6). Then we obtain

$$\int_t^{t+1} \int_D \varphi(\tau - t) \partial_y \psi(z) \gamma^\pm(\tau, z) \, dz \, d\tau - \int_t^{t+1} \int_D \varphi'(\tau - t) \psi(z) \gamma^\pm(\tau, z) \, dz \, d\tau$$

$$+ \int_t^{t+1} \varphi(\tau - t) \int_D (\nabla u^\pm(\tau, z) \nabla \psi(z)) \, dz \, d\tau = 0.$$

Changing variables $\sigma = \tau - t$, passing to the limit $t \to \infty$, and using the assumptions concerning φ we obtain

$$\int_D (\nabla u_\infty^\pm(z) + e\gamma_\infty^\pm(z)) \nabla \psi(z) \, dz = 0$$

for all $\psi \in V$.

It remains to show

$$u_\infty^\pm \geqslant 0, \quad 0 \leqslant \gamma_\infty^\pm \leqslant 1, \quad u_\infty^\pm (1 - \gamma_\infty^\pm) = 0 \qquad \text{a.e. in } D,$$

but this follows from the pointwise convergence (see (4.3), (4.4) and (4.1)). Then using the same arguments as in [BR], Proof of Theorem 1, we have

$$\int_D (\nabla u_\infty^\pm + e\gamma_\infty^\pm) \nabla v \leqslant 0$$

for all $v \in H^1(D)$, $v \geqslant 0$ on $\Gamma_0 \cap \{G = 0\}$, $v = 0$ on $\Gamma_0 \cap \{G > 0\}$.

Now the statement of Theorem 1.8 can be shown as follows. $u_\infty^\pm$ are solutions of the stationary problem 1.2. Because of the assumption (1.7) we have $u_\infty^+ = u_\infty^-$. On account of $u^- \leqslant u \leqslant u^+$ a.e. on D_∞ (see (4.1)) and $u^\pm(t, \cdot) \to u_\infty^\pm$ in $L^p(D)$, $1 \leqslant p < \infty$ (see (4.4)) we get

$$u(t, \cdot) \to u_\infty := u_\infty^\pm \quad \text{in } L^p(D).$$

This proves Theorem 1.8.   □

References

[A1] H.W. Alt, *Stömungen durch inhomogene poröse Medien mit freiem Rand*, Journal für die reine und angewandte Mathematik **305** (1979), 89–115.

[A2] H.W. Alt, *The fluid flow through porous media. Regularity of the free surface*, Manuscripta math. **21** (1977), 255–272.

[A3] H.W. Alt, *A free boundary problem associated with the flow of ground water*, Arch. Rat. Mech. Anal. **64** (1977), 111–126.

[AG] H.W. Alt and G. Gilardi, *The behaviour of the free boundary for the dam problem*, A. Fasano, M. Primicerio, (ed.) Free boundary problems: theory and applications **1** (1983), 69–76, Montecatini.

[AL] H.W. Alt and S. Luckhaus, *Quasilinear elliptic-parabolic differential equations*, Math. Z. **183** (1983), 311–341.

[BE1] J. Bear, *Dynamics of fluids in porous media*, "American Elsevìer", New York, 1972.

[BR] H. Brezis, *The dam problem*, A. Fasano, M. Primicerio, (ed.) Free boundary problems: theory and applications **1** (1983), 77–87, Montecatini.

[CC] J. Carillo-Menendiz and M. Chipot, *On the uniqueness of the solution of the dam problem*, A. Fasano, M. Primicerio, (ed.) Free boundary problems: theory and applications **1** (1983), 88–104, Montecatini.

[FB] A. Friedman and E. Di Benedetto, *Periodic behaviour for the evolutionary dam problem and related free boundary problems*, Preprint (to appear).

[GI] G. Gilardi, *A new approach to evolution free boundary problems*, Comm. in Partial Differential Equations **4(10)** (1979), 1099–1122.

[KE] N. Kenmochi, *Asymptotic behaviour of the solution to a Stefan-type problem with obstacle on the fixed boundary*, Hiroshima Math. J. (to appear).

[K] D. Kröner, *Asymptotic behaviour of the solution of the non-steady dam problem for incompressible fluids*, Control and Cybernetics **14** (1985), 247–273.

[KR] D. Kröner and J.F. Rodrigues, *Global behaviour of a porous media equation of elliptic-parabolic type*, J. Math. pures et appl. **64** (1985), 105–120.

[R] J.F. Rodrigues, *On the free boundary of the evolution dam problem*, A. Fasano, M. Primicerio, (ed.) Free boundary problems: theory and applications **1** (1983), 125–134, Montecatini.

[TO1] A. Torelli, *Su un problema di frontiera di evoluzione*, Boll. U.M.I. (4) **11** (1975), 559–570.

New Results Concerning the Singular Solutions of the Capillarity Equation

Marie-Françoise Bidaut-Veron

1. Introduction

In this work we study the global existence and uniqueness of a singular solution of the capillarity equation in $\mathbb{R}^N$:

$$\text{div}(Dv/\sqrt{1 + |Dv|^2}) = \kappa v, \tag{1}$$

with a $\kappa < 0$. Our results concern symmetric solutions with an isolated singularity at the origin. Then equation (1) takes the equivalent form in $]0, +\infty[$:

$$(r^{N-1}u'/\sqrt{1 + u'^2})'(r) = -(N-1)r^{N-1}u(r), \tag{2}$$

where $u(r) = \sqrt{-\kappa(N-1)^{-1}}v(\sqrt{-(N-1)\kappa^{-1}}r)$.

In [2], P. Concus and R. Finn proved the existence of a singular solution U of (2) with the following development near 0:

$$U(r) = -\frac{1}{r}\left(1 - \frac{N+3}{2(N-1)}r^4 + r^4\varepsilon(r)\right), \tag{3}$$

with $\lim_{r\to 0}\varepsilon(r) = 0$. They got the local uniqueness of solutions of that form, such that $\varepsilon(r)/r^p (p < 4)$ and $\varepsilon'(r)$ are bounded. Moreover rU has an asymptotic development in powers of r^4 but the formal Taylor series is divergent. They conjectured that U could be continued for all positive r and that the global solution was unique, up to the change of U to $-U$. In [3] they proved that any singular solution of (2), negative near the origin, was asymptotic to U when r goes to 0.

Let us notice also that the solution U is concave near the origin, since from (2) and (3),

$$\frac{U''(r)}{(1 + U'^2(r))^{3/2}} = -(N-1)\left(U(r) + \frac{U'(r)}{r\sqrt{1 + U'^2(r)}}\right) = -2r^3(1 + o(r)). \tag{4}$$

Here we first prove the global existence of singular solutions u of (2). The idea is to formulate an equivalent problem for a new unknown function:

$$z(r) = \frac{u'(r)}{\sqrt{1 + u'^2(r)}};$$

this leads us to a semilinear second order equation of the form

$$\Delta z(r) = \Phi(z(r), r), \tag{5}$$

where $r \mapsto \Phi(r, z(r))$ is not defined for $r = 0$ and may oscillate for small r. With an *a priori* energy estimate for z we prove that any local solution can be extended as a global solution.

Then we prove the uniqueness of a singular solution u of (2), negative and concave near the origin. For that aim, in a first step, we improve the results of local uniqueness: with a fixed point method analogous to [2], applied to function z, we find a larger class of uniqueness for functions z and u; namely, we get the local uniqueness of singular solutions u of (2) such that $(u'(r) - (1/r^2))$ is not too large. Then in a second step, as maximum principle fails in equation (5), we use local comparison methods to get some accurate estimates for z for small r, when u is negative and concave near the origin. From these estimates we deduce that such functions z and u belong to the new class of uniqueness defined in step one. For details we refer to [1].

2. Global Existence

First we formulate the problem in terms of function z and prove easily the following assertion.

Lemma 1. *The existence and uniqueness of a C^2 function u, singular solution of (2), negative near the origin, is equivalent to the existence and uniqueness of a C^2 function z, solution of the following equation*

$$\Delta z(r) = z''(r) + (N - 1)\frac{z'(r)}{r}$$

$$= \Phi(z(r), r) = (N - 1)z(r)\left(\frac{1}{r^2} - \frac{1}{\sqrt{1 - z^2(r)}}\right), \tag{6}$$

with limit conditions

$$\lim_{r \to 0} z(r) = 1, \qquad \lim_{r \to 0} z'(r) = 0. \tag{7}$$

Remarks. Functions z and u are linked by the relations

$$z(r) = u'(r)/\sqrt{1 + u'^2(r)} = \sin \psi(r), \tag{8}$$

$$z'(r) + (N - 1)\frac{z(r)}{r} = -(N - 1)u(r), \tag{9}$$

where ψ is the angle between the tangent at $(r, u(r))$ and the r-axis. To obtain equation (6) we derive equation (9) and write (8) in the equivalent form

$$u'(r) = z(r)/\sqrt{1 - z^2(r)}. \tag{10}$$

We get the limit conditions with the estimation proved in [3] for small r:

$$z(r) \geqq 1 - \frac{(\pi + \sqrt{2})^2}{2} r^4 + o(r^4). \tag{11}$$

Notice also that (9) is equivalent to

$$z(r) = -\frac{N-1}{r^{N-1}} \int_0^r \rho^{N-1} u(\rho) \, d\rho. \tag{12}$$

Finally, the variations of z and the concavity of u are linked by the obvious relation

$$z'(r) = u''(r)/(1 + u'^2(r))^{3/2}. \tag{13}$$

The form of equation (6) shows the importance of the sign of the difference $(\sqrt{1 - z^2(r)} - r^2)$. Now we give an essential estimate of the energy for the function z, obtained by multiplication by $z'(r)$ in equation (6) and integration.

Lemma 2. *Let z be a solution of problem* (6), (7) *on an interval* $[0, R[$. *Then*

$$E(r) = \frac{z'^2(r)}{2(N-1)} + \frac{1 - z^2(r)}{2r^2} - \sqrt{1 - z^2(r)}$$

$$= \frac{1}{2} \left(\frac{z'^2(r)}{N-1} + \left(\frac{\sqrt{1 - z^2(r)}}{r} - r \right)^2 - r^2 \right) < 0, \tag{14}$$

and $E'r) < 0$ in $]0, R[$. *Consequently*

$$0 < \sqrt{1 - z^2(r)} < 2r^2, \tag{15}$$

$$|z'(r)| < \sqrt{N - 1} \min(r, \sqrt{2}), \qquad in \ ]0, R[. \tag{16}$$

Then we can prove the global existence theorem.

Theorem 1. *Each solution z of* (6), (7), *likewise each singular solution u of* (2), *has a unique extension defined on* $]0, +\infty[$.

OUTLINE OF THE PROOF. We show with Lemma 2 that z, hence also u, have a unique extension, defined on an interval $[0, R_m)$, and that R_m cannot be finite. In particular the function U can be extended to $[0, +\infty[$ and we note Z the function z associated to U by relation (8). $\qquad\qquad\square$

Remarks. In [1] we make precise the behavior near infinity of any solution z or u. We first use the estimate of the energy to prove that z, z', u, u' go to 0 when r goes to infinity. Then we compare z near infinity to the solutions of the linear equation

$$\Delta\zeta(r) = \zeta''(r) + (N-1)\frac{\zeta'(r)}{r} = (N-1)\zeta(r)\left(\frac{1}{r^2}-1\right), \tag{17}$$

namely the Bessel equation of order 1 when $N = 2$; with the Sturm–Liouville theorem, we prove that z and u are oscillatory: the function z admits a countable number $(r_n)_{n\geq 1}$ of zeros, asymptotically separated by a distance of $\pi/\sqrt{N-1}$, the function u admits a countable number $(\rho_n)_{n\geq 1}$ of zeros, such that

$$0 < \rho_1 < r_1 < \rho_2 < r_2 < \cdots < \rho_n < r_n < \rho_{n+1} < r_{n+1} < \cdots,$$

and has a unique extremum in r_n on $[\rho_n, \rho_{n+1}]$; moreover $(|u(r_n)|)_{n\geq 1}$ is decreasing, so as $(|z'(r_n)|)_{n\geq 1}$. We can also give lower and upper bounds for ρ_1 and r_1. All those properties may be compared to the results of [4], [6] concerning nonsingular solutions.

3. Uniqueness Under Growth Conditions

The estimates of ε and ε' for the solution U of (2) defined by (3) show that the function Z associated to U satisfies

$$Z(r) = 1 - \frac{r^4}{2} + O(r^8), \tag{18}$$

near the origin.

First we improve the results of local existence and uniqueness of [2].

Theorem 2. *Let* $M < M_0 = (N+8)/3\sqrt{N-1}$. *Then the function* Z *is the unique solution* z *of problem* (6), (7) *such that the following holds near the origin:*

$$z(r) = 1 - \frac{r^4}{2} + r^6 w(r), \qquad |w(r)| \leq M. \tag{19}$$

In the same way the function U *is the unique singular solution* u *of* (2) *such that, near the origin,*

$$u'(r) = \frac{1}{r^2} + \omega(r), \qquad |\omega(r)| \leq M. \tag{20}$$

OUTLINE OF THE PROOF. We use very accurately a fixed point method introduced in [2] to get local existence and uniqueness of a solution z of (6), (7) in the largest class of functions possible. Let z be a C^2 function, solution of problem (6), (7), with $w(r) = r^6(z(r) - 1 + (r^4/2))$ such that $|w(r)| \leq M$ in an interval $]0, R[$. Let $g(r) = \sqrt{N-1}\,r^{(N+8/2)}w(r)$; then

$$r^3 g''(r) + 3r^2 g'(r) + (N-1)g(r) = \sqrt{N-1}\,r^{N+2}F(w(r),r), \tag{21}$$

where

$$F(w,r) = 2(N+2)r^2 + \frac{N(N-4)}{4}r^4 w + (N-1)w$$

$$+ \phi\left(1 - \frac{r^4}{2} + r^6 w, r\right). \tag{22}$$

Hence with limit conditions (7), introducing the kernel of equation (21), we deduce that

$$w = T(w), \tag{23}$$

where T is the mapping from $B_{M,R} = \{v \in C^\circ([0,R]) | \max_{r \in [0,R]} |v(r)| \leq M\}$ to the space $C^\circ([0,R])$ of continuous functions on $[0,R]$, defined by

$$T(w)(r) = \frac{r^{-(N+8)/2}}{\sqrt{N-1}} \int_0^r \tau^{(N+2)/2} F(w(\tau), \tau) \sin\frac{\sqrt{N-1}}{2}\left(\frac{1}{\tau^2} - \frac{1}{r^2}\right) d\tau. \tag{24}$$

Conversely, any solution w of (23) in $B_{M,R}$ is a C^∞ function in $]0,R[$, and $z(r) = 1 - (r^4/2) + r^6 w(r)$ is a solution of problem (6), (7) in $]0,R[$. Then we prove that for $M < M_0$ and sufficiently small R, one of the iterates of the mapping T is a strict contraction, hence T has a unique fixed point. Hence we get the uniqueness for z from Theorem 1, and for u from (8) and (10). $\square$

Remark. For any function z, the estimate (19), near the origin, with an $M < M_0$ is equivalent to the estimate

$$|\sqrt{1 - z^2(r)} - r^2| \leq \tilde{M} r^4, \tag{25}$$

near the origin, with an $\tilde{M} < M_0$. In order to get the uniqueness, we have to prove (25). In the following, we prove it for solutions z which are nonincreasing near the origin.

Theorem 3. *The function Z is the unique solution z of (6), (7) such that z is nonincreasing near the origin. The function U is the unique singular solution u of (2) such that u is concave near the origin.*

OUTLINE OF THE PROOF. We compare such a solution z in the neighborhood of any small $\rho > 0$ with a function y of the form

$$y(r) = ar^2 + br + cr^{1-N} \qquad (a, b, c \in \mathbb{R}), \tag{26}$$

such that

$$y(\rho) = z(\rho), \quad y'(\rho) = z'(\rho), \quad y''(\rho) = z''(\rho). \tag{27}$$

First we study the differences $(y(\rho \pm \rho^2) - z(\rho))$, use the fact that $z'(\rho) = O(\rho)$ from (16), and deduce an estimate of the form

$$\sqrt{1 - z^2(\rho)} = \rho^2 + O(\rho^3). \tag{28}$$

Then we consider differences $(y(\rho \pm \rho^3) - z(\rho))$, at the points ρ where the quotient $(\sqrt{1 - z^2(\rho)} - \rho^2)/\rho^4$ may have a local extremum, for which $z'(\rho) = O(\rho^3)$, and get a more accurate estimate:

$$|\sqrt{1 - z^2(\rho)} - \rho^2| \leq 4 \sqrt{\frac{2}{3(N-1)}} \rho^4 + o(\rho^4), \tag{29}$$

at those points and then at any small $\rho > 0$.

Now let us observe that (29) has the form (25), with an $\tilde{M} < M_0$, so that, from Theorem 2, z is equal to Z. As z is nonincreasing if and only if u is concave, from (13), we deduce the analogous result for u.

We conjecture that one can obtain uniqueness results for singular solutions u of (2) and solutions z of (6), (7) without any growth condition.

References

[1] M.F. Bidaut-Veron, *Global existence and uniqueness results for singular solutions of the capillarity equation*, Pacific Jl. Math (to appear).

[2] P. Concus and R. Finn, *A singular solution of the capillarity equation*, I: *Existence*, Invent. Math. **29** (1975), 143–148.

[3] P. Concus and R. Finn, *A singular solution of the capillarity equation*, II: *Uniqueness*, Invent. Math. **29** (1975), 149–160.

[4] P. Concus and R. Finn, *The shape of a pendent liquid drop*, Philos. Trans. Roy. Soc. London A **292** (1979), 307–340.

[5] R. Finn, *Some properties of capillarity free surfaces*, Seminar on Minimal Submanifolds, Princeton Univ. Press (1983), 323–337.

[6] R. Finn, *On the pendent liquid drop*, Z. Anal. Anwendungen **4** (1985), 331–339.

Continuous and Discontinuous Disappearance of Capillary Surfaces

Paul Concus and Robert Finn

1

We consider the problem of finding a capillary surface $u(x)$ in a cylinder Z with section Ω, in the absence of gravity. The surface is to meet Z in (constant) contact angle γ and project simply onto Ω. It is known that to each Ω there exists a critical angle $\gamma_0 \in [0, \pi/2]$ such that a surface exists if $\gamma_0 < \gamma \leqslant \pi/2$, while no surface exists if $0 \leqslant \gamma < \gamma_0$. We show here that if $0 < \gamma_0 < \pi/2$ and if Ω is smooth, then there is no surface at $\gamma = \gamma_0$, while if Ω has corners a surface can in some cases be found. In the former case, the surface disappears "continuously" and always becomes unbounded in a sub-domain Ω^* of positive measure, as $\gamma \searrow \gamma_0$. In the latter case the surface can be bounded, and even analytic with the exception of the corner points. Some applications of the result are given and the exceptional case $\gamma_0 = 0$ is discussed.

2

In the presence of a gravity field the free surface $u(x)$ in a capillary tube of section Ω is determined as a solution of the governing equations under very general conditions on the form of Ω (see e.g. [1, 2, 3]). If the gravity field is removed, there may be no surface satisfying the physical requirements, even for convex analytic Ω (see [4, 5, 6]).

In the absence of gravity, the governing equations take the form

$$\operatorname{div} Tu = \frac{\Sigma}{\Omega} \cos \gamma, \qquad Tu = \frac{Du}{\sqrt{1 + |Du|^2}} \tag{1}$$

in Ω, and

$$v \cdot Tu = \cos \gamma, \tag{2}$$

on $\Sigma = \partial\Omega$. Here γ is the prescribed boundary angle, v is unit exterior normal on Σ. We use the symbols, $\Sigma, \Omega, \ldots$ to denote alternatively a set or its measure. We may assume $0 \leqslant \gamma \leqslant \pi/2$ (see, e.g. [4]).

The material of this paper derives from the following condition for existence (see [9, Theorem 7.1]).

Suppose that an inequality

$$\int_\Sigma |f| \, ds \leqslant \mu \int_\Omega |Df| + \Upsilon \int_\Omega |f| \tag{3}$$

holds for every f of bounded variation, $f \in BV(\Omega)$, with $\mu < 1/\cos\gamma$, $\Upsilon = $ const depending only on Ω. Suppose further that an isoperimetric inequality

$$\min(\Omega^*, \Omega\backslash\Omega^*) < C(\Omega \cap \partial\Omega^*)^2 \tag{4}$$

holds for every Cacciopoli set $\Omega^* \subset \Omega$. Then a solution of the capillary problem $(1, 2)$ *exists if and only if the functional*

$$\Phi[\Omega^*; \gamma] \equiv \Gamma - \Sigma^* \cos\gamma + \left(\frac{\Sigma}{\Omega}\cos\gamma\right)\Omega^* \tag{5}$$

satisfies $\Phi > 0$ for every Cacciopoli set $\Omega^* \subset \Omega$ with $\Omega^* \neq \varnothing, \Omega$.

Here

$$\Gamma = \int_\Omega |D\phi|, \qquad \Sigma^* = \int_\Sigma \phi \, ds,$$

where ϕ is the characteristic function of Ω^*; in the second integral, ϕ is taken to be the trace on Σ^* from within Ω (see [7, ch. 1]). When it exists, the solution is real analytic and bounded in Ω, satisfies (1) strictly and (2) in a generalized sense, and is unique up to an additive constant among all solutions in that class, in particular among all functions that satisfy $(1, 2)$ strictly.

An inequality of the form (3) appears first in Emmer [1], who proved it for a Lipschitz domain with $\mu = \sqrt{1 + L^2}$, $L = $ Lipschitz constant. Thus (3) holds for any smooth domain for any $\mu > 1$. An inequality (4) holds in considerable generality, and certainly for smooth domains (cf. [7, p. 25]). Thus, we conclude from the above theorem and from the form of (5) that to every smooth Ω there corresponds $\gamma_0 \in [0, \pi/2]$ such that existence holds if $\gamma_0 < \gamma \leqslant \pi/2$, whereas existence fails if $0 \leqslant \gamma < \gamma_0$. Simple examples show that the case $0 < \gamma_0 < \pi/2$ occurs commonly; that is, it must be expected that solutions of $(1, 2)$ will exist for some values of γ in the admissible physical range, but not for others.

3

We ask what happens at $\gamma = \gamma_0$ and find a result that depends in an unusual way on the smoothness of Ω.

Theorem. *Let Ω be such that $0 < \gamma_0 < \pi/2$. Suppose* (3) *holds with $\mu < 1/\cos\gamma_0$ and* (4) *also holds. Then the zero-gravity capillary problem* (1, 2) *for Ω admits no solution when $\gamma = \gamma_0$.*

PROOF. From the condition for existence given in Sec. 2, we find that if $\gamma < \gamma_0$ then there exists $\Omega^* \subset \Omega$ with $\Phi[\Omega^*; \gamma] \leqslant 0$, $\Omega^* \neq \varnothing$, Ω. Taking now a sequence $\gamma_j \nearrow \gamma_0$, we find that the Ω_j^* form a family of sets of bounded perimeter in Ω, hence there is a subsequence that converges to Ω_0^*, in the sense that the characteristic functions converge in $L^1(\Omega)$.

Suppose $\Omega_0^* = \varnothing$. We choose f in (3) to be the characteristic function ϕ_j of Ω_j^*, obtaining

$$\Phi[\Omega_j^*; \gamma_j] \geqslant (1 - \mu\cos\gamma_j)\Gamma - C\Omega_j^*$$

for some fixed constant C. By (4) we have, since $\Omega_j^* \to \varnothing$, that $\Omega_j^* < C\sqrt{\Omega_j^*}\,\Gamma$ for all sufficiently large j. Since $\mu < 1/\cos\gamma_0$ there follows $\Phi[\Omega_j^*; \gamma_j] > 0$ for large enough j, which is a contradiction.

Suppose $\Omega_0^* = \Omega$. We introduce the "adjoint" functional

$$\Psi[\Omega^*; \gamma] \equiv \Gamma + \Sigma^*\cos\gamma - \left(\frac{\Sigma}{\Omega}\cos\gamma\right)\Omega^*$$

and observe that $\Phi[\Omega^*; \gamma] = \Psi[\Omega\backslash\Omega^*; \gamma]$. Using (4), we thus find

$$\Phi[\Omega_j^*; \gamma] \geqslant (1 - C\sqrt{\Omega\backslash\Omega_j^*})\Gamma > 0$$

for large enough j, again a contradiction.

Thus $\Omega_0^* \neq \varnothing$, Ω. We now use an observation that first appears in Gerhardt [8]. We set $\beta_0 = \cos\gamma_0$, $\beta_j = \cos\gamma_j$ and consider

$$\Phi_0 - \Phi_j = \int_\Omega |D\phi_0| - \int_\Omega |D\phi_j| - \beta_0 \int_\Sigma (\phi_0 - \phi_j)\,ds + (\beta_j - \beta_0)\int_\Sigma \phi_j\,ds$$

$$+ \frac{\Sigma}{\Omega}\beta_j \int_\Omega (\phi_0 - \phi_j)\,dx - \frac{\Sigma}{\Omega}(\beta_j - \beta_0)\int_\Omega \phi_0\,dx.$$

Let $A_\delta \subset \Omega$ denote a strip of width δ adjacent to Σ. Let $\eta(x) \in C^\infty(\bar{\Omega})$, with $\eta(x) \equiv 1$ on Σ, $\eta(x) \equiv 0$ in $\Omega_\delta = \Omega\backslash A_\delta$, $0 \leqslant \eta \leqslant 1$ in Ω. Since $\beta_j \to \beta_0$, $0 < \beta_j < 1$, and $\phi_j \to \phi_0$ in $L^1(\Omega)$, we may write

$$\Phi_0 - \Phi_j \leqslant \int_{\Omega_\delta} |D\phi_0| - \int_{\Omega_\delta} |D\phi_j| + \int_{A_\delta} |D\phi_0| - \int_{A_\delta} |D\phi_j|$$

$$+ \beta_0 \int_\Sigma |(\phi_0 - \phi_j)|\,ds + \varepsilon_j,$$

where $\varepsilon_j \to 0$ with j. To the integral over Σ we apply (3), obtaining

$$\beta_0 \int_\Sigma |(\phi_0 - \phi_j)\eta|\,ds \leqslant (1 - \varepsilon)\int_{A_\delta} |D\phi_0| + (1 - \varepsilon)\int_{A_\delta} |D\phi_j|$$

$$+ C(\Omega; \delta)\int_\Omega |\phi_0 - \phi_j|\,dx$$

for some $\varepsilon > 0$. We have further

$$\limsup_j \left\{ \int_{\Omega_\delta} |D\phi_0| - \int_{\Omega_\delta} |D\phi_j| \right\} \leqslant 0$$

by the semicontinuity of the length integral (cf [7], Theorem 1.9). Thus

$$\limsup_j \{\Phi_0 - \Phi_j\} \leqslant (2 - \varepsilon) \int_{A_\delta} |D\phi_0|.$$

Since δ is arbitrary, we find

$$\Phi[\Omega_0; \gamma_0] \leqslant \limsup_j \Phi[\Omega_j; \gamma_j] \leqslant 0,$$

and since $\Omega_0 \neq \varnothing$, Ω, the existence condition in Sec. 1 yields that no solution exists at γ_0. The theorem is proved.   □

Since $0 < \cos \gamma_0 < 1$, we conclude immediately from the theorem and from the remarks preceding it that for a smooth domain there is no surface of the type sought for the critical angle γ_0. It is remarkable that if Ω is allowed to contain corners, then a solution at γ_0 can in some cases be found. A simple example is obtained by choosing for Ω a regular polygon with interior corner angle 2α. A lower hemisphere whose equatorial circle circumscribes Ω then provides an explicit (bounded) solution of (1, 2), analytic except at the corner, and with $\gamma = \pi/2 - \alpha$. It is shown in [4] that in the case of an interior boundary angle 2α no solution can exist if $\gamma < \pi/2 - \alpha$. Thus in the example considered, $\gamma_0 = \pi/2 - \alpha$ and a solution exists at γ_0.

In such a situation the solution is seen to disappear discontinuously as γ decreases through γ_0. In the former (nonexistence) case, e.g., for smooth boundaries, we may regard the disappearance as continuous, in the sense that each γ for which existence holds lies in a γ-interval of existence.

The theorem is sharp, in the sense that the result can fail when $\mu = 1/\cos \gamma_0$; that can be seen from the above example of the regular polygon. Nevertheless, the result can hold in much more generality than the case of Lipschitz continuity that seems to be indicated. For example, an inward cusp with opening angle 2π can be admissible; see, e.g., the discussion in Chapter 6 of [9]. Precise geometric conditions for admissibility have not yet been established.

$$4$$

As an illustration of continuous disappearance, consider the configuration of Figure 1, bounded by line segments L_1, L_2 and by tangent circular arcs C_1, C_2 of radius 1, ρ respectively. It can be shown that if h is sufficiently large then there is a unique γ_0 and circular arc C_0 meeting the boundary Σ in angle γ_0 (as in Figure 1), such that $\Phi[\Omega^*; \gamma_0] = 0$. If $\gamma > \gamma_0$ a solution of (1, 2) exists, while if $\gamma \leqslant \gamma_0$ there is no solution. As $\gamma \searrow \gamma_0$ the solutions $u(x; \gamma)$ can be

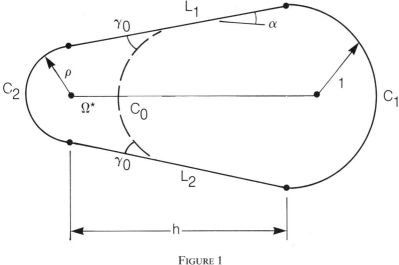

FIGURE 1

normalized to converge to a solution $u^0(x)$ throughout $\Omega\backslash\Omega^*$ and to infinity throughout Ω^*.

In the above example, for given ρ there holds $\lim_{\alpha\to 0}\gamma_0 = \pi/2$. Thus, the more "parallel" the segments L_1, L_2 are to each other, the smaller is the range of γ for existence. Nevertheless, if $\alpha = 0$ a solution exists for any γ, regardless of h. This singular behavior can be seen in the curves of Figure 2 relating γ_0 to ρ for differing values of α. (The computed tabular points for the curves are spaced with increment 0.025 in ρ and connected with straight line segments.) As $\alpha\to 0$ the curves tend to the upper and right hand boundary segments; however $\alpha = 0$ yields only the indicated single point $(1,0)$.

5

The varied behavior that can occur in the case $\gamma = 0$ is illustrated further by the example of Figure 3, in which C_1 is a semicircle of radius 1 tangent to the two parallel line segments, and C_2 is a circular arc of radius ρ.

One can show that there exists a unique $\rho_0 = 1.974\ldots$ such that if $\rho < \rho_0$ a solution surface exists for $\gamma = 0$ (and, _a fortiori_, for all $\gamma > 0$) for any value of h. If $\rho = \rho_0$ then an "extremal" Ω_0^*, for which $\Phi[\Omega_0^*;0] = 0$, is obtained by inscribing a semicircular arc of radius 1 at any point in the strip as indicated. Thus, no solution can exist at $\gamma_0 = 0$. If $\gamma > 0$, one can show that $\Phi[\Omega^*,\gamma] > 0$ for any possible "extemal" Ω^*, regardless of h, so that a solution exists for this case. One can show, as for the example of Sec. 4, that as $\gamma\searrow 0$ the solution $u(x;\gamma)$ can be normalized to converge to a solution $u^0(x)$ throughout $\Omega\backslash\Omega^*$, and to infinity throughout the Ω^* that is determined in the strip by the arc C_0.

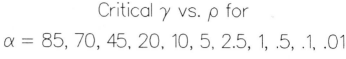

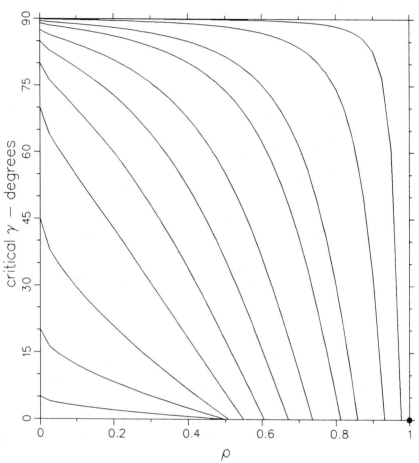

FIGURE 2

6

An elaboration of the above configuration yields the example indicated in Figure 4. Here $\delta = 60°$, ABC is a semicircle; AF, CG are circular arcs of unit radius.

If $\rho = 1.974$, calculations show that for $h = 10$ a solution exists if $\gamma > \gamma_0 \approx$ 13.76°. At $\gamma = \gamma_0$ an "extremal" arc AC with $\Phi = 0$ appears, of radius $\rho_1 \approx$ 1.028. A sequence u^j of solutions corresponding to data $\gamma_j \searrow \gamma_0$ can be nor-

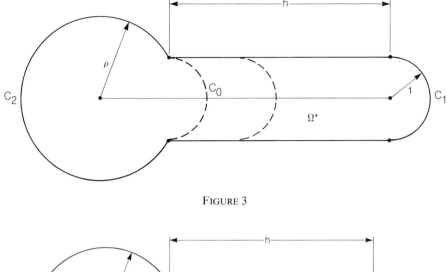

FIGURE 3

FIGURE 4

malized so that u^j tends to a solution to the left of AC, and to infinity to the right of AC.

If $\rho = 1.984$, a solution exists for $h = 10$ if $\gamma > \gamma_0 \approx 15.25°$. At this angle an "extremal" arc DE with $\Phi = 0$ appears, of radius $\rho_2 \approx 1.039$. A sequence u^j of solutions corresponding to data $\gamma_j \searrow \gamma_0$ can be normalized so that u^j tends to a solution to the left of DE, and to infinity throughout the region to the right of DE.

Thus, the region in which the solution becomes infinite can be made to shift, essentially discontinuously, with small changes in the domain and data.

Acknowledgments. We wish to thank E. Giusti for comments that have led to an improvement of our original result, with simpler proof. This work was supported in part by NASA grant NAG3-146, by NSF grant MCS83-07826, and by the Applied Mathematical Sciences subprogram of the Office of Energy Research, U.S. Department of Energy under contract DE-AC03-76SF00098.

References

[1] M. Emmer, *Esistenza, unicità e regolarità delle superfici di equilibrio nei capillari,* Ann. Univ. Ferrara **18** (1983), 79–94.

[2] R. Finn and C. Gerhardt, *The internal sphere condition and the capillary problem.* Ann. Mat. Pura Appl. **112** (1977), 13–31.

[3] E. Giusti, *Boundary value problems for non parametric surfaces of prescribed mean curvature,* Ann. Scuola Norm. Sup. Pisa **3** (1976), 501–548.

[4] P. Concus and R. Finn, *On capillary free surfaces in the absence of gravity,* Acta Math. **132** (1974), 177–198.

[5] R. Finn, *Existence criteria for capillary free surfaces without gravity,* Indiana Univ. Math. J. **32** (1983), 439–460.

[6] R. Finn, *A subsidiary variational problem and existence criteria for capillary surfaces,* J. Reine Angew. Math. **353** (1984), 196–214.

[7] E. Giusti, *Minimal Surfaces and Functions of Bounded Variation,* Monographs in Mathematics, vol. 80, Birkhäuser, Boston (1984).

[8] C. Gerhardt, *Global regularity of the solution to the capillary problem:* Ann. Scuola Norm. Sup. Pisa **3** (1976), 157–175.

[9] R. Finn, *Equilibrium Capillary Surfaces,* Springer-Verlag (to appear).